SOTHIS PRESS

Sothis Press aims to completely reimagine the public domain book with an emphasis on universal access to captivatingly designed print editions. We believe that technology will enable every single book ever published to be perpetually accessible in beautiful print form. Sothis is beauty in ubiquity.

www.sothis.press
1 Waterhouse Square, London, ECIN 2ST

Hot Water Supply and Kitchen Boiler Connections

A Text Book on the Installation of Hot Water Service in Residences and Other Buildings and Methods of Connecting Range Boilers, Steam and Gas Water Heaters

William Hutton

SOTHIS PRESS

ISBN 978-1-0368-0181-6

Hot Water Supply and Kitchen Boiler Connections is in the public domain.

This edition is published by Sothis Press © 2023. The company is registered in England and Wales. The Company Number is 14900389.

Hot Water Supply and Kitchen Boiler Connections

A Text Book on the Installation of Hot Water Service in Residences and Other Buildings and Methods of Connecting Range Boilers, Steam and Gas Water Heaters

By William Hutton

Based on Articles from
Metal Worker, Plumber & Steam Fitter
with Addenda and Useful Tables

New York
David Williams Company

T
H8

PREFACE.

Every man who has been engaged in the plumbing business or who has had to do with the design and construction of buildings requiring a supply of hot water at the various sanitary fixtures, will concede that there is no other branch of building construction in which trouble is easier to find by a departure from correct design or by improper construction.

Those who do not admit it need only study the columns of trade papers devoted to plumbing topics to find that more problems are presented for solution in this line than in any other branch of plumbing, and the same can be said of all countries possessing trade papers. There need be no reflection on the plumber in admitting this. The subject is one that requires more study of principles than practical experience in construction, and it is generally found that when unsatisfactory systems are constructed the mistake is in design through an improper understanding of principle, and not because of poor workmanship.

The popular hand-book "Kitchen Boiler Connections"—dealt principally with piping problems and with the connections to boilers in the smaller type of residence. While the examples shown in the book which is designed to replace it cover the larger buildings as well, it has been recognized that the former are the more important by reason of their far greater number and therefore the examples of piping construction and connections for small buildings are shown in greater variety. This is considered all the more necessary, as the opportunity for departure from certain standard types of construction are more in small buildings than in large ones, as a rule, owing to greater variety in architectural design.

All of the methods of connecting heating appliances of various types have come under the author's personal observation, and it has been his intention to show as nearly as possible such connections as may be considered standard and which are likely to become necessary at some time in the practice of others, while eliminating examples which might be considered freakish or exceptional. Much of the material in the book has appeared in the pages of *"Metal Worker, Plumber and Steam Fitter,"* and such parts of the previous book on this subject as applied to up-to-date practice have been retained.

While it is not to be expected that every combination that can be satisfactorily used is shown, it is hoped that the examples which have been selected are varied enough to serve the purpose of guiding the inexperienced mechanic to the selection of a form of construction which will give satisfactory service for the special conditions he may have to work to, and it is hoped that these have been set forth in such a manner that a little study will enable him to grasp the principles which have to be kept in mind in selecting them. If this is done and each problem carefully considered with these principles in mind, there will be less need for the services of the "trouble man" in hot water installations.

WM. HUTTON.

TABLE OF CONTENTS.

CHAPTER I.

Principles of Heating, Combustion, Transmission of Heat.

Is there any single branch of the plumbing business which provides so much food for discussion, so many knotty problems to solve or so many possibilities of failure as that branch of work which comes under the heading of Hot Water Supply and Kitchen Boiler Connections?

Take up the current number of any one of the trade journals and an enquiry will be found from some member of the craft who is "up against it." Look back over a year of your own experience and see if you cannot recollect some instances when you found it very hard to decide on the proper, or at least the most advantageous, method of installing a hot-water generator or system of distribution.

It is precisely this choice of methods, the uncertainty of gaining the results expected and the difficulty of deciding which of several plans is best suited to a particular problem that lead to the troubles experienced by most plumbers at one time or another, and in many cases to loss of patronage through unsatisfactory working of the apparatus he has installed. In practically all of the other branches of our business, the methods are well defined, and the results of a failure to comply with established practice so obvious that difficulties are encountered only by the less experienced or incompetent men.

But in this business of Hot-Water Supply, it is not only the young mechanic, but often the old and presumably much experienced craftsman who "falls down" on some point or other with the attendant failure to secure satisfactory results from his work. And the only possible explanation is lack of comprehension of the principles governing the movement of water in a heating and distributing system.

The purpose of this book is to present to such as are interested in the subject a few of the chief factors which make for success, or failure, in a hot-water installation, a few of the facts known to most experienced men, but possibly new to the young journeyman or apprentice, and examples of the application of these principles and methods which have come under the writer's observation or been submitted to him for opinion by others from time to time.

For the benefit of the new boy who wants to know "how the hot water gets into the boiler, anyhow," it is necessary to explain what causes circulation.

Circulation.

Circulation, as understood in its application to a hot-water heating or domestic hot-water supply system, is the movement of bodies of water from the primary heating appliance, which may be a kitchen stove, a regular hot-water heater, an automatic gas or a steam-heating contrivance, to the storage tank or boiler, and the return of water from thence to the source of heat again. It is also applied to the movement of hot water through a circuit of pipe to some point near the fixtures to be supplied with hot water for heating purposes or for bathing and similar purposes, and its return from that point to the tank or boiler to be reheated and again returned to the point of discharge for immediate service when wanted.

What causes circulation? Briefly, the difference in density or weight per given volume of two columns of water at different temperatures.

And now to explain the meaning of Density. Density has been defined as the "Ratio of mass to volume."

Water has a maximum density at a temperature of about 40 deg. F. That means that a given quantity of water will weigh more at 40 deg. than at 35 deg. or at 110 deg., for a rise in temperature causes the molecules to expand in volume, thereby taking up more space, with the obvious result that our "given quantity" is larger, while its weight has remained as it was. A decrease in temperature has the same effect, but at 32 deg. the water solidifies and takes the form of ice.

We take advantage of this natural law in causing circulation to take place in heating systems, and its application to practical problems is as follows:

When heat is applied to a vessel containing water, the molecules at the point to which heat is applied expand and, being lighter for the same space occupied than the cooler particles surrounding them, are pushed upward; those taking their place pass through an identical process, follow, and immediately a constant upward current is established.

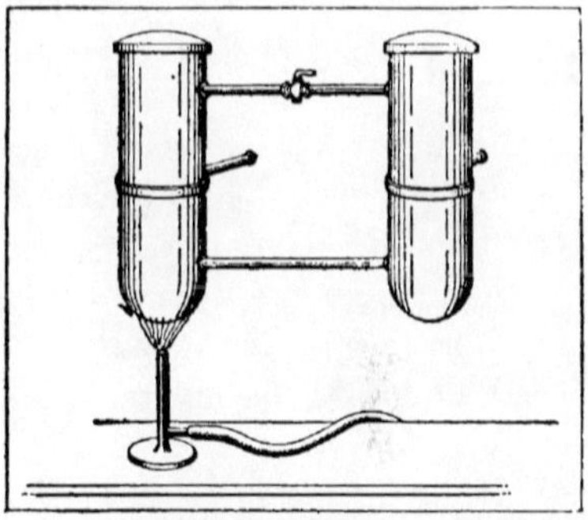

Fig. 1. Appliance Used to Demonstrate Circulation of Water.

Suppose, then, we take two vessels and connect them together with two tubes, one close to the bottom and one near the top of the vessels. We now have an apparatus which represents approximately the usual heater and boiler combination.

Apply heat to the lower part of the vessel, representing the heater, and the local circulation just described begins immediately. Very soon the water will begin to pass through the upper tube into the second vessel, while a corresponding quantity will pass from that vessel through the lower tube into the first, where it will become heated and rise until it reaches the upper part, then through the tube back into the lower part of the original vessel again.

Thus complete circulation has been established and will be maintained as long as heat is applied. It will be obvious that the temperature will not be the same in all parts of the vessel, owing to the cooling effect of its walls, but the local circulation induced through this is so slight as to be almost imperceptible and has no effect on the circulation through the tubes to the other vessel.

The force that causes the water in a hot-water heating or supply system to flow is sometimes termed the "Motive Column," and is the pressure due to the head of water gained by expansion over the head at the temperature before heating. It must always be borne in mind that hot water will move, or "circulate," only when there is a heavier or cooler body of water to displace it, and the motive force is proportional to the difference in temperatures of the ascending and descending columns and also to the height. Thus the motive force in a circuit 30 ft. high is twice as great as in one 15 ft. high, and this is the reason why circulation to a radiator or other fixture on an upper floor is generally faster than to some other on the same or on a lower level.

But even with a considerable difference in temperature in the two columns, and a considerable height of circuit, the force is very slight and circulation comparatively easily held back or entirely stopped.

A pitch in the wrong direction, causing a pocket where air may form, is often enough to entirely prevent satisfactory circulation and horizontal runs of circulation pipe must be carefully laid so that these pockets are impossible.

If lead pipe is used every precaution to prevent sagging must be taken or trouble will surely come to the fitter.

We have given some consideration to the matter of "Circulation" and its cause. Let us enumerate and try to define some of the properties and natural forces which we bring into operation in promoting it, or which may act to retard it, in any system of hot-water heating or supply.

Heat.

Heat is a phenomenon hard to define.

Briefly: Heat is not matter, but a form of energy requiring matter to act through. It has no weight. The addition of heat to any substance, liquid or solid, makes no difference whatever to its total weight, although it may change its nature to a very considerable degree.

Heat is measured by a standard unit known as the British thermal unit, or shortly, B.t.u., which is the amount of heat required to raise the temperature of 1 lb. of water at 39 deg. 1 deg. F. and also by a standard termed a calorie, which is roughly equal to 4 B.t.u.

It cannot be destroyed or created; it can simply be transmitted from one substance to another, from one form of matter to another form, or transformed into work from which it may be again recovered.

The temperature at which oxygen goes into rapid combustion—that is, causes what we call fire to take place—differs with different bodies. Thus phosphorus ignites at 150 degrees F., sulphur at 480 degrees, while the hydrocarbons require a temperature of nearly 1000 degrees to kindle them.

Development of Heat and Fire.

Fire is visible heat or light. Luminous heat, light, is motion; heat is motion. The more intense the motion the greater the heat and light. Whenever motion is arrested heat is liberated, and whenever motion is set in action heat is expended. The same amount of heat that it took to start the train is given out when the train is stopped, whether it be slowly or suddenly. If it be suddenly, the heat is more intense, brakes are very much heated, and sparks are produced at the point of friction.

The heat of ordinary combustion is generated by the impact of the particles or atoms of oxygen with those of the combustible. The greater the affinity of oxygen for any substances or the greater the facilities for their union the more intense the action and the greater the heat.

The kindling of a fire is an art that has grown with the intelligence of the race. First, the rubbing of two sticks; then the flint and tinder, and last the phosphorized match—the perfection of convenience. In each instance the cause of the fire is the same, friction or arrested motion. Phosphorus has the lowest ignition point of any substance that would be safe to use, a slight stroke producing sufficient heat to ignite it. After ignition the energetic union of the elements is sufficient to maintain the fire. Two requirements, therefore, are necessary for a fire. First, elevation of temperature to the ignition point; second, a supply of air to the fuel.

On the other hand, only two things are necessary to extinguish it—lower the temperature or cut off the supply of air. The best results are obtained when both expedients can be employed. Chemical extinguishers are mostly carbon dioxide generators, and their extinguishing properties are due more to the water projected at a lower temperature and at the base of the flame than to the gas, although it is a perfect extinguisher, could it be provided in sufficient quantity economically.

The Production of Heat by Combustion.

Combustion may be defined as a rapid chemical combination, resulting in heat and light. The combining elements are: (a) Oxygen, which is usually derived from atmospheric air; (b) either carbon or hydrogen, or a compound of the two. Sulphur sometimes appears with carbon and hydrogen, and also combines with oxygen. The substance that is formed by the chemical union is called the product of combustion; and the heat that is produced by the combustion of a unit weight (1 lb.) of the fuel is called the heat of combustion.

When hydrogen burns it combines with the oxygen of the air and forms water and gives off 62,000 heat units per pound of hydrogen burned. When carbon is burned one atom of carbon may take up one atom of oxygen, forming carbon monoxide and developing 4,400 B.t.u. per pound. When the combustion process is carried to its limit; in other words, when two atoms of oxygen combine with one atom of carbon, carbon dioxide is formed, developing 14,500 B.t.u. per pound of carbon. The difference in the amount of heat developed shows how important it is to have a sufficient draft and supply of air.

Heat may be transmitted to the water in a heating system from the fire-box or gas burners in three different ways—by radiation, by convection and by conduction.

Radiation.

Radiant heat possesses the remarkable property of passing through the atmosphere without perceptibly raising its temperature. These heat rays, whether emanating from a glowing or dark hot body, are transmitted through the air at right angles to the point of emanation until they are absorbed or reflected by another body; in this particular case the walls of the fire-box, the exposed portions of the boiler or the coils.

Radiant heat passes away from hot-water pipes, boilers, radiators, etc., so long as the temperature of these is higher than that of any objects within its sphere of influence. The amount of radiant heat emitted and received varies with the nature of the surface affected, but in the same surface the power of absorption and radiation is equal. Different metals have different powers of radiation, and it is important to bear this in mind in designing or installing heating appliances.

Convection.

The conductivity of liquids for heat is very slight. Nevertheless, heat is rapidly transferred throughout their volume, owing to their qualities of expansion and mobility, by direct transport of the heated particles. In describing the action set up by heat and the cause of circulation, it was shown that the particles expanding on the application of heat, and becoming lighter, volume for volume, than those surrounding them, were rapidly carried upward and displaced by the heavier and cooler particles in regular sequence.

These particles in their upward passage transmit a portion of their heat by convection, until in the regular rotation of heating, expanding, rising, loss by convection, sinking and reheating, a complete circulation is maintained, and eventually a more or less uniform temperature obtained through the whole body of water in the system.

Heat applied near the bottom of any vessel or apparatus designed for the purpose of heating water is therefore the most effective; as the particles being driven away from contact with the heated bottom surface, rise through the longest section of cold water in the boiler, transferring a proportionately larger amount of heat by convection to the descending particles. That applied at the sides, as in the side flues of a hot-water boiler, causes the currents of heated particles to start only from that point where the heat is applied; while heat applied at the top is almost useless except as a means of checking loss by radiation from the boiler or its connections.

Conduction.

The power of transmission of heat by conduction varies in different metals to a considerable extent, and a knowledge of the relative coefficiencies is of considerable value, as it enables the mechanic to select the material best suited to prevent waste and diffusion of heat, or to convey it with the best hopes of successful results.

The expansion of various metals with a rise in temperature must always be kept in mind and an acquaintance with the relative amount of expansion of various metals is important. Tables showing coefficiencies of expansion and conductivity are available and are likely to be of considerable assistance to the student in obtaining reliable design.

Imperfect Combustion.

From the foregoing data it will be seen that imperfect combustion of fuel is not only a nuisance through the production of excessive smoke but is really wasteful. This explains the economy shown in the operation of a stove or water heater with a well designed firebox and grate. If the supply of oxygen is sufficient to effect the higher and more complete combustion the total heat available from 1 lb. of anthracite may be 14,500 B.t.u., but making due allowance for the loss in flues and by radiation to the plates of the heater and so to the air a total transmission of 8,000 heat units to the water in the coil or water front may be looked for. Under the same conditions one authority has calculated the heat available from soft or bituminous coal as 6,500 B.t.u., from coke as 9,350 B.t.u., from hickory wood as 4,300 B.t.u. from 1 cu. ft. of coal gas as 650 B.t.u. and from natural gas 950 B.t.u.

These are the figures that may be taken as a basis on which to estimate heating capacities of fireboxes or heating surfaces, allowances having been made for losses in stoves of average design and construction.

To secure satisfactory combustion of fuel in the ordinary kitchen stove or tank heater the oxygen supplied should be in sufficient quantity to reach all parts of the fuel. If properly proportioned the two atoms of oxygen necessary to combine with each atom of carbon and hydrogen in the fuel will be provided and not only will there be an absence of smoke but the efficiency of the heater will be higher.

Strains and Stresses.

The various strains and stresses in materials such as metals engendered by expansion or by methods of construction of the appliances the student is interested in should also be the subject of his consideration, as he will be better able to judge of the efficiency of his work, if he has some knowledge of the breaking or tensile strengths and the stresses that can safely be withstood.

Stresses may be classified as follows: "Tensile," or pulling, "compressive," "transverse," or bending, "shearing" and "torsional," or twisting stress.

Strain is deformation, or change of shape of a body as the result of a stress.

Elasticity is the power a body has of returning to its original form after a stress on it is withdrawn.

Elastic limit is the unit stress under which the body becomes permanently strained or deformed.

In the practice of hot-water pipe fitting these terms are met with frequently. For example:

Lead pipe has little elasticity. Under the stresses set up by repeated contraction and expansion, it becomes strained or permanently elongated, causing sagging; or where movement of the body is prevented, as at a junction with another pipe, the low tensile strength of the material causes it to rupture by repeated stress and distortion of the component particles at some particular point.

Then there are the bursting and shearing stresses and strengths, as exemplified in the amount of pressure a boiler will stand without a failure of its plates or pulling apart from its rivets, or that various kinds of pipe will withstand under heavy pressures or sudden shocks.

CHAPTER II.

Corrosion of Water Fronts, Boilers and Pipes.

Most plumbers who have had an opportunity of observing the effect of corrosion in pipes have noticed the fact that pipes conveying hot water have been more affected than those in which cold water is conveyed. While this has been known in the trade for a long time it is only within recent years that any attempt has been made to ascertain the conditions which would tend to hasten the process with the object of avoiding them if that is at all possible.

In this case corrosion must not be confounded with sedimentation or precipitation of insoluble matter.

In all waters brought from lakes or open reservoirs there is as a rule a fair proportion of matter in suspension which will be precipitated on boiling and the same is true as regards water impregnated with lime and other impurities of a mineral nature. It is a common thing to find galvanized iron and even brass or lead pipe which has become coated with a deposit from the water which has gradually built in on the walls of the pipe until its bore has become almost closed. This cannot be called corrosion as there has been no deterioration of the pipe, no eating away and weakening of its walls. Occasionally, however, this latter condition becomes evident and this is what is referred to as true corrosion and to be seen more frequently in hot water pipes than in cold. Investigations of the cause of rapid corrosion of hot water pipes in recent years have brought out the statement that rusting tests at varying temperatures show that corrosion is most active between 140 and 180 deg. Fahr. and that above that temperature there is comparatively little action on the pipe.

This has been explained as being due to excess of oxygen. Pure water at normal pressure will dissolve 14.7 parts per million of oxygen at 32 deg. Fahr. and 7.60 parts per million at 86 deg. Fahr. At 210 deg. oxygen is practically insoluble so that as the water is heated the solubility at normal pressure becomes less. In the conditions in which water is heated for domestic hot water supply the oxygen must remain in solution owing to the fact that the water is heated under pressure and in a closed vessel. Thus the water may be said to be supersaturated with oxygen and judging from the effects noted in different tests its rapid passage over the surface of iron or steel pipes in circulation through the system hastens corrosion entirely because of the presence of an excessive supply of oxygen in a condition favorable for this action.

The obvious remedy for this state is to reduce the temperature of the water passing through the pipes and where this has been possible and has been put into effect the results are said to have been satisfactory. In ordinary kitchen range water heating systems, however, this is not so easily arranged. A temperature of only 110 deg. at the water front would necessitate a larger storage tank than is easily found room for and even with that it would be hard to circulate the water quickly enough to prevent its reaching a much higher temperature when a strong fire was maintained in the range. Where a tank heater is employed this latter objection does not apply and it is possible by using a heater of ample proportions together with a storage tank of more than the actual size required under the present conditions to maintain a larger supply of water at the lower temperature and so obtain a satisfactory service while reducing the effect of corrosion considerably. Where the water front must be used a compromise can be effected by using brass pipe for the connections between boiler and water front as the average temperature in the boiler will seldom rise above 110 and the water passing into the secondary circulation will therefore have a less harmful effect on the piping than that circulating between the water front and the boiler where the temperature is higher.

Stoppage of Pipes and Water Fronts by Lime.

Filling of pipes and water front passages by sediment deposited as the result of precipitation of matter in solution, such as sulphates and carbonates of lime, is frequent in districts where the water supply is obtained from springs or wells drilled in rock of a limestone or chalky formation. Such water is always hard, its hardness being measured by degrees based upon the amount of carbonates or sulphates the water contains. One degree of hardness would thus mean that one gallon of water contained one grain of carbonate of lime or chalk and as every grain of chalk in the water will curdle eight grains of soap before a lather is produced the degrees of hardness in the water may be very accurately estimated by a careful test using soap as the medium.

Waters which contain sulphates of lime are classed as permanently hard and their mineral contents are not precipitated to any extent by boiling. If, however, the hardness be caused by the presence of carbonates then the water is classed as temporarily hard and the chalk will be deposited in any vessel in which the water is boiled vigorously. This is the white and hard deposit which builds up on the walls of pipes between the water front and the boiler, which chokes up the water ways of the water front also and which manifests its presence by poor heating and by snapping sounds in the tank. It can be softened to some extent before it enters the water front by adding to the water a little lime water, which will absorb the carbonic acid which holds the salts in solution and so precipitate both the lime in the water added and that in the water. A modification of this process is seen in the water softening appliances known as injectors, shown in Fig. 2, and which are intended to be placed in the return connections of water fronts so that all water passing through them will be subjected to a treatment which will neutralize the action of the heat upon the contents of the water.

The deposit of the carbonates will also be accompanied by the precipitation of a certain amount of calcic sulphate should that be present. When water of known hardness must be used for domestic purposes the means of heating should be such that the water will never reach boiling point. This can be effected either by the use of a tank heater and storage tank of such generous proportions that a low fire may be run and the supply of water make up in quantity what is lost in temperature or by shielding the surface of the water front or coil which is exposed to the fire by inserting a brick in front of it and by using large pipes with easy bends so that the water will circulate freely and will have little tendency to deposit any matter in suspension at the elbows or sharp turns. The effect of such deposits on the plates of a water front or the pipes in a coil has already been commented upon and the danger that may arise from this building up process is too great to neglect.

Removal of Lime From Waterbacks.

Where it is known to exist the water front and coil connections should be made in such a manner that they can be readily

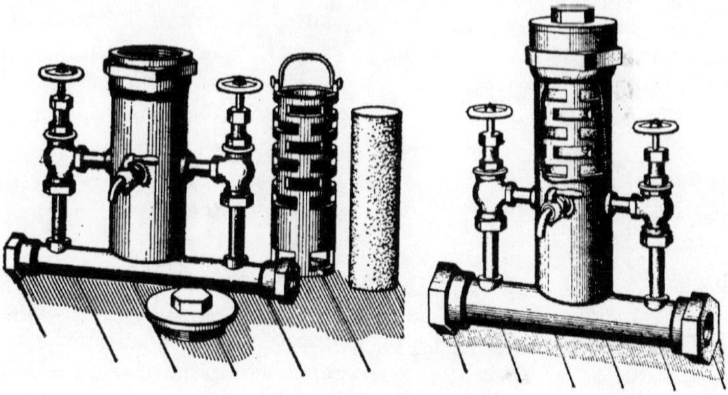

Fig. 2. Water Softening Apparatus.
The Component Parts. The Parts Assembled.

removed and the deposit cleaned out at frequent intervals. There are several methods of removing a deposit of this character among which may be mentioned that of pouring into the water front a solution of one part hydrochloric acid and five parts water, which has been recommended by many who have had occasion to try it. The water front should be heated until the solution boils gently, when the water back is to be removed from the range and the ordinary impurities washed out by being thoroughly flushed. Then the solution can be poured in and the water back placed on top of the stove or any other place where it can be brought to a boiling point. The incrustation, it is said, will be dissolved within the course of an hour or two and can be flushed out with water. If there is a thick incrustation and the first treatment fails to remove it the treatment should be repeated until it takes effect.

Instead of using the acid a solution of common washing soda may be tried. This is commonly used for the removal of scale from boilers and according to the nature of the deposit and the amount of it may give satisfactory results in water fronts also. The water front should be allowed to soak in the solution and the process of removing the incrustation will be hastened by heating. The incrustation should leave the metal and when dry form a powder which is not difficult to remove. When the deposit is very heavy the water front should be allowed to soak in it for several days and an extra quantity of soda should be used.

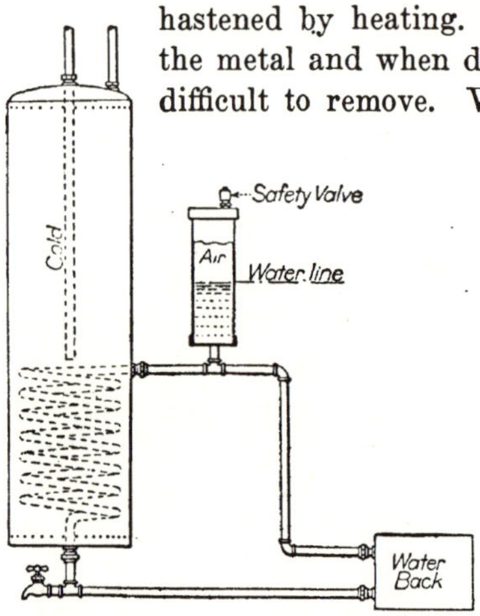

Fig. 3. Method of Connecting Boiler Using Alkaline Water.

Another method of avoiding the precipitation of the mineral contents of water in water fronts, pipes and boilers is that illustrated in Fig. 3. This is said to have been very satisfactory in many places where the water is of an alkaline nature and where the usual method of piping the range connections is not satisfactory through the excessive deposit which is made in the system and which necessitates constant changing of water fronts and cleaning of pipes and boilers. From the illustration it will be seen that the flow pipe from the water front does not connect with the boiler proper, but with a coil that is placed inside it. This coil is kept filled with pure rain water from a small reservoir shown connected to the upper pipe connection to the water front.

If the reservoir is covered at the top and a safety valve fitted a pressure can be maintained on the coil which will allow of carrying a much higher temperature than would be possible were it open, while the evaporation will be much reduced and thus necessitate less frequent filling. One of the boilers ordinarily used for heating water by steam is quite suitable for connecting to the water front in this manner as it requires about 1 sq. ft. of heating surface for each 5 gall. of water heated. The reservoir may be made out of 3 in. iron pipe about 18 in. long, the top being closed by a cap drilled and tapped to fit the safety valve and having a plug for easy filling as often as is necessary. While this system is one requiring attention it is efficacious as the alkaline water in the boiler which is fed from the main pipe in the usual manner is heated by conduction and does not become heated to the high temperature that it would when passed through a water front or coil in the firebox in the usual manner. Thus the precipitation of the mineral contents is almost entirely avoided as there is very little effect on the water until boiling point has been reached. It is a good plan when using water that carries sediment or that is likely to precipitate lime to use tees at all bends or turns instead of ells and to plug one outlet. Thus the pipes can be easily cleaned with a wire when the sediment begins to fill up the pipe.

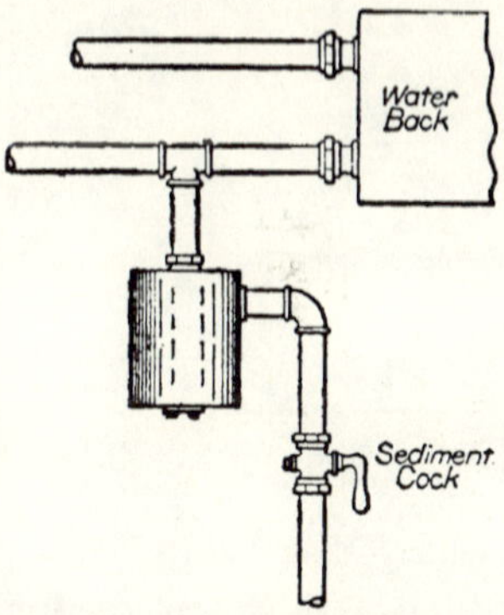

Fig. 4. A Sediment Chamber.

Sediment Collecting Chambers.

A sediment chamber which collects mud and other matter precipitated in the heating system is shown at Fig. 4. A clean-out screw on the bottom allows of the easy removal of the sediment and its flushing with a hose.

This appliance can be fitted on the return connection of the water front or immediately under the lower tapping of the boiler as found desirable and convenient.

A sediment chamber of a little different design and intended to be fitted to the lower tapping of a vertical boiler is shown in Fig. 5. From the illustration it will be seen than an inner tube extends through the connection into the boiler, thus keeping the inlet to the water front above the level at which sediment will be likely to stand. A hose bibb or a pipe connection with a valve screwed into the outer chamber enables the sediment to be flushed out at intervals.

Heat Losses From Boilers and Pipes.

While the heat losses from an ordinary kitchen boiler with a connection from a water front heated by a coal fire may be comparatively unimportant, there are other conditions which make such losses a matter worth consideration. In tanks of larger capacity than those used in the average household, such losses by radiation become important and the usual method of avoiding them is to cover the tank with a non-conducting covering of asbestos or some such substance. Smaller boilers may have a covering of similar nature if desired and special coverings of sheet asbestos and canvas are made which can be readily laced into place with hooks provided for the purpose. Another method is to use a lagging of closely fitted boards bound with metal bands. Where the boiler is heated by gas and a thermostatic control of the valves is used such provisions are economical, as the constant radiation from the boiler when uncovered will cool off the contents and cause the gas valve to open more frequently than the ordinary demand of the building would call for. Where such coverings are dispensed with, however, it is possible to retard to some extent the radiation by using the proper paint to finish the boiler with and a knowledge of the effects of different pigments and bronzes on the radiation from substances coated with them is of some value in this connection and may be applied with profit.

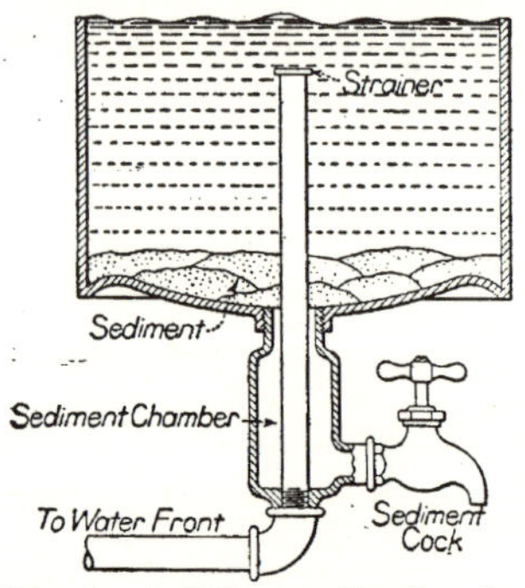

Fig. 5. A Sediment Chamber for a Kitchen Boiler.

In an extensive series of tests made several years ago it was found that the use of the bronzes, paints and enamels commonly used for painting radiators and other metal surfaces affected the rate of transmission of heat from the surfaces by as much as 27 per cent. Applying two coats of copper bronze reduced the heat transmission from a radiator by 26 per cent. while the addition of two coats of terra cotta enamel on the top of the bronze not only reversed this effect but gave a higher rate of transmission than the bare metal gave. The effect of adding these two coats was to increase the transmission of heat 28.6 above the transmission when the bronze only was used and to check this two more coats of copper bronze were applied on the top of the terra cotta enamel when the conducting capacity of the surface fell 26.5 per cent., bringing back the rate of transmission to about the same as it was with the two coats of copper bronze. Following out the tests it was found that the loss of heat was nearly the same with 14 coats of paint as it was with 2 coats and that the effect depended apparently on the nature of the last coat applied. The tests also showed that the color of the paints and enamels applied had some effect on the rate of transmission and the final conclusion was that copper or aluminum bronze offered the greatest resistance to the passage

HEAT LOSSES IN B.t.u. Per sq. ft.		B.t.u radiated per hr. per sq. ft. of surface per deg. diff. in temp.	Average room temp.	No. of test	Efficiency
1. Rad. plain as received from factory	These paints of two Coats each were painted over one another in the order given.	2.82	74.4	1	0.997
2. Rad. plain as received from factory		2.925	76 0	2	1.005
3. Rad. painted with copper bronze		2.835	63.1	3	0.761
4. Rad. painted with copper bronze		2 78	72.3	4	0.752
5. Rad. painted with terra cotta enamel		2.78	74.5	5	1.038
6. Rad. painted with copper bronze		3.05	66.3	6	0 735
7. Rad. painted with light brown varnish		2.74	74.1	7	0.977
8. Rad. painted with oak brown varnish		2.74	72.9	8	0.977
9. Rad. painted with aluminum bronze		2.70	71.8	9	0.730
10. Rad painted with aluminum bronze	This series follows one another.	2.77	70 5	10	0.724
11. Rad. painted with silver gray enamel		2.77	66.7	11	0.970
12. Rad. painted with snow-white enamel		2.72	67.6	12	1.01
13. Rad. painted with bronze green enamel		2.67	64.2	13	0.997
14 Rad. painted with no luster green enamel		2.67	.0	14	0.956
15. Rad. painted with maroon gloss japan		2.63	6	15	0.997
16. Rad. painted with shellac and copper bronze powder		2.61	5	16	0.850
17. Rad. painted with copper bronze powder and linseed oil		2.66	.0	17	0.760
18. Rad. painted with white paint	Painted over one another	2.66	9	18	0.987
19. Rad. painted with terra cotta paint		2.64	64.4	19	1.00
20 Rad. painted with light green paint		2.62	[illegible].8	20	0.989
21. Rad. painted with light green paint zinc		2.72	77.2	21	1.00
22. Rad. painted with terra cotta paint zinc		2.77	77.7	22	0.964
23. Rad. painted with white paint zinc		2 68	76.0	23	1.01

Table of Heat Losses Through Painted Iron Surfaces.

of heat and therefore is the best material that can be used to paint boilers or pipes which are in such a position that it is not practicable to cover them with the ordinary form of covering.

The foregoing table shows the rate of heat transmission per sq. ft. per deg. difference in temperature in a cast iron radiator with steam at an average temperature of 224 deg., and the losses from a boiler containing water at 180 to 200 deg. Fahr. may be estimated from this, making due allowance for the difference in outside and inside temperatures. Thus the rate of transmission would be less when the room temperature was higher, but the table will give results nearly enough correct for all practical purposes.

CHAPTER III.

Water Fronts, Coils and Heaters.

The usual provision made in kitchen range construction for water heating purposes is that of a water front or water back. This is simply a hollow cast iron box which is designed to take up the space afforded by one or more sides of the firebox and which receives its designation of water front or water back according to the position in which it is intended to be fitted. In the smaller sizes of ranges intended for use in cottages or small apartments in connection with a boiler of 30 or 40 gall. capacity the water front is invariably of the pattern occupying the space at the front or back of the fire box. In the larger ranges with ovens at each side of the fire box two sides or three may be occupied by it, affording a much larger heating surface which is capable of supplying hot water in sufficient quantity, in combination with a boiler of 60 to 100 gall. capacity, for a large household.

The style of the range determines the position of the flow and return tappings in the water front. Those shown in the accompanying illustrations, Figs. 6 to 10, show all the positions in common use. From these it will be seen that the outlets may be at the side of the range or the back according to what the design requires, but the relative position of flow and return connections are as a rule the same—the flow as near the top as possible and the return as close to the under side of the casting as can be arranged. This prevents the accumulation of steam in the water front with attendant noise and even danger, and also admits of sediment being well scoured out when the sediment cock at boiler is opened.

Frequently there is a partition in the water front extending nearly the full length. This assures a good circulation through it and a better heating effect as the water has to pass through a longer circuit in contact with the heating surface than it might take without this provision. The dotted lines in the illustrations show how air or steam may accumulate in the water front when the upper connection is made at too low a point and how sediment may accumulate below the level of the lower pipe. The danger from the accumulation of air arises from the fact that the metal may become overheated in contact with a fierce fire and the interior pressure will then bulge it outward, weakening the walls to the extent that a rupture may occur. The sudden liberation of water at a high temperature and pressure may result in its flashing into steam, which will do great damage when the right conditions exist. The same results may follow the accumulation of an undue amount of sediment on the inner surface of the water front. In this case the walls become overheated by reason of the extra thickness, the deposit of foreign matter prevents the conduction of heat by the water and allows the metal to become red hot and in its weakened state to give way under the pressure. Indication of the presence of sediment in dangerous amount is given as a rule by a swelling or bulge on the wall in contact with the fire which grows larger on repeated heating to high temperatures.

Fig. 6

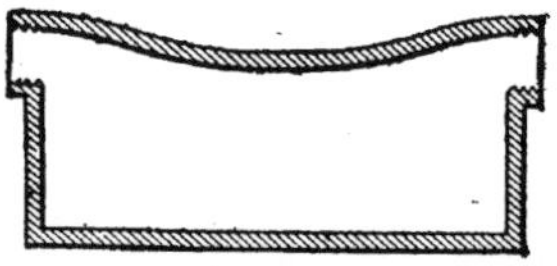

Fig. 8.

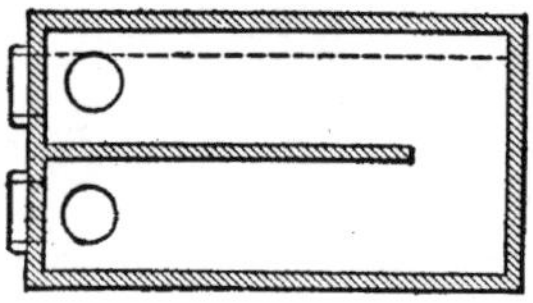

Fig. 7.

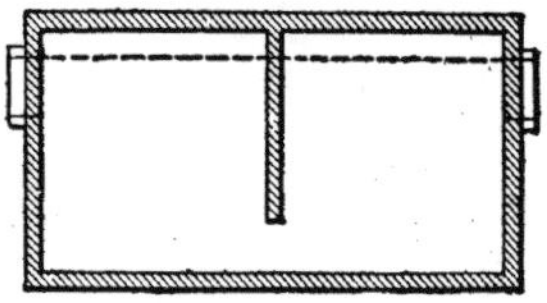

Fig. 9.

Section Through Water Backs.
Dotted Lines Show How Air May Collect Above Connections.

Some water fronts have the partition in a vertical position. This is convenient with some designs of ranges as it enables the flow and return connections to be made at the same level, the partition deflecting the water inside so as to afford a circulation. Theoretically the position at the back of the fire box should be the most effective in which to place the water heater as the hot gases passing to the flues will then pass over a portion of it but in practice the difference between front and back or sides is not noticeable. What is of more importance is the depth and width of the fire box. When the size of the fire box is reduced too much in either direction it will be found that the heat transmission from the fuel to the water front is so great as to affect the baking qualities of the range seriously unless the flue to which the stove is connected possesses an unusually good draft. Occasionally also this condition works in the opposite direction and the fuel in contact with the water front will be noticeably dull and the water supply unsatisfactory. It will also be found hard to main-

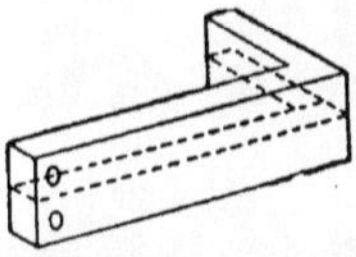

Fig. 10. Section Through 2-Side Water Back.

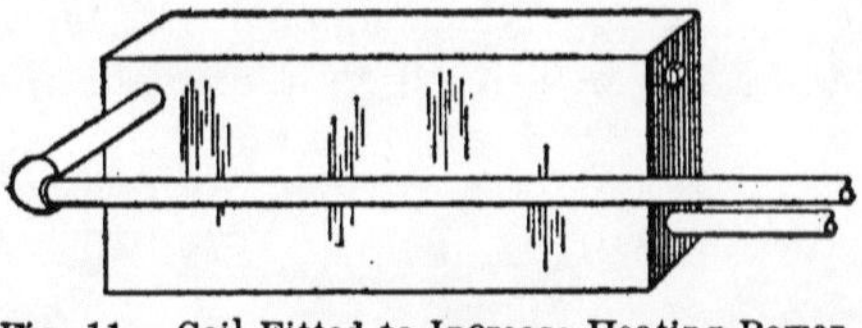

Fig. 11. Coil Fitted to Increase Heating Power.

tain a fire when the dampers are checked down. The reason is the obvious one that the range is overtaxed and therefore the size of the fire box should be a factor in deciding as to the practicability of heating a sufficient supply of water for domestic purposes when it is proposed to use a water front in a range in which none has been previously used.

Heating Large Quantities of Water.

In large establishments using powerful cooking stoves and requiring large amounts of hot water it is quite possible and in fact common to install a boiler of 60 to 80 gall. capacity which is heated by a water back in the range. If the double oven type of range is used which has the fire box in the center a water back which extends around three sides of the fire box may be provided. This will afford a large heating surface which with the strong fire usually run where a large amount of cooking is done will be sufficient to heat the water satisfactorily. It is quite possible to expose 200 sq. in. of heating surface with a water back of this description and on the basis of 2½ sq. in. to the gallon of water that is sufficient to heat 80 gallons with ease.

It must not be overlooked however that such a large water back is a severe tax upon a fire and when overdone will lead to unsatisfactory baking in the ovens. Therefore a water front such as that shown in Fig. 10 which is carried round two sides of the fire box only and which even then is capable of heating a more than ordinarily large boiler may be the safer to use in the majority of cases. When placing water fronts in position it is always well to bed them with fireclay and to point the joints with that or with stove putty. This will add considerably to the heating value of the fire by preventing an accumulation of ash in the crevices and by preventing a current of air from passing behind the water front and bricks.

Another point to look out for is that the stove is standing level. It is not uncommon in old houses to find floors so much out of the level by sinking of walls that the water front is high

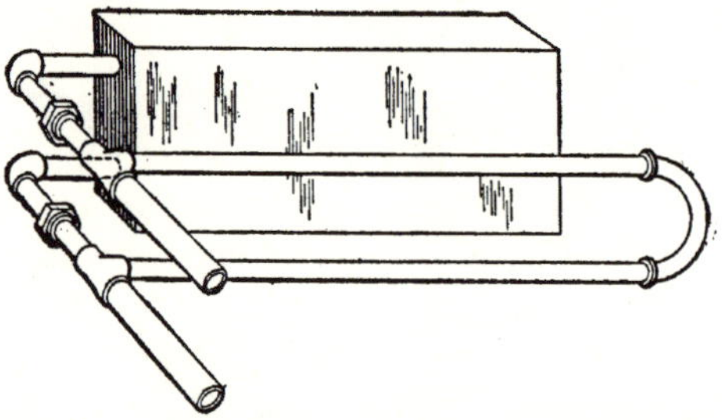

Fig. 12. A Combination of Coil and Water Front.

at one end and this may considerably affect the circulation as well as act injuriously on the water front should air or steam collect therein. It also throws the tappings of the water front out of the correct line so that the pipes connecting to it will be pitching either up or down more than is desirable. This latter condition has sometimes to be overcome by cutting a crooked thread on the connections as it may be also due to a tapping not being correctly made.

When an additional supply of hot water is required through the extension of the plumbing system in a house or for any other reason it is often accomplished by tapping the water front as shown in Fig. 11 and carrying a brass pipe around the fire box in contact with the hot coals or exposed to the hot gases passing over the top of the oven to the flues. If the additional requirements are not great a pipe passing round two sides and so through the stove plates may be entirely satisfactory, or it can be carried along the top of the oven and returned to the fire box end again before passing out to be connected to the boiler. It is important to see that a pitch upward is maintained, otherwise the circulation will be impeded and accompanied by rumbling sounds caused by the accumulation of air or steam in the coil.

Another method of extending the heating surface of a water front is shown in Fig. 12. This is really a combination of coil and water front, the coil being carried along the back of the fire box and connected into the flow and return pipes to boiler as shown. Such a coil is easily constructed of brass pipe with one return bend, unions being inserted between the

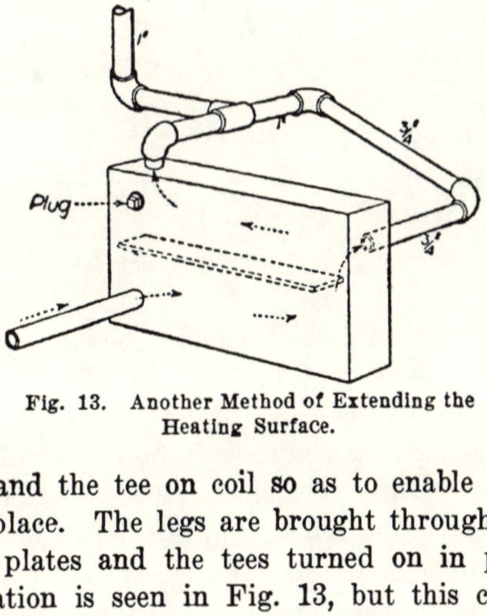

Fig. 13. Another Method of Extending the Heating Surface.

water front and the tee on coil so as to enable the coil to be pushed into place. The legs are brought through holes drilled in the stove plates and the tees turned on in position. Yet another variation is seen in Fig. 13, but this calls for drilling the water front in another place. This ensures that all air is removed from the water front and when the stove is of such construction that the connection can be made in this manner the additional heating power that such a coil affords will be considerable. The arrows in the illustration denote the path of the water in circulating throught the water front. The tapping for the coil connection is made just above the level of the partition in the water front and that from the top is taken off at one end in a line with the return connection. The flow connection used before the extension was made is plugged. In piping up a water front in this manner it is imperative that there should be no burrs in the pipe or fittings and a smoother working job will be obtained if there is room in the fire box to make the piece of pipe and the tee of 1-in. pipe as shown, as then there will be no retarding of the flow by friction with consequent over-heating and noisy operation.

Instead of a cast iron water back it is common to use a coil made of iron or preferably of brass pipe. This is generally built by the plumber to fit the stove installed and consists of the

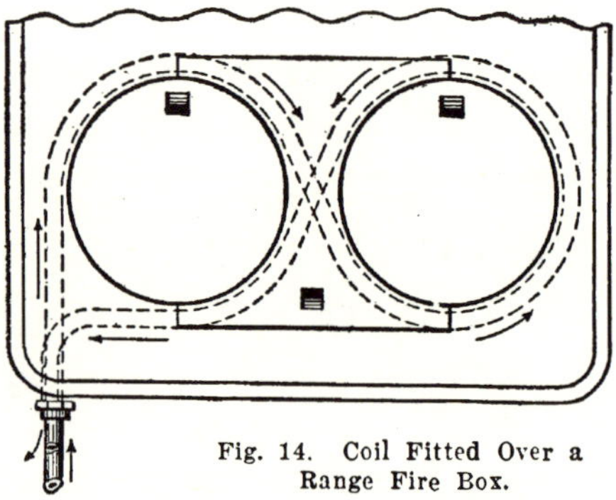

Fig. 14. Coil Fitted Over a Range Fire Box.

ordinary elbows and return bends in common use. It can be made to fit one, two, three or four sides as desired and is generally built in a simple two pipe style, although with a fire box of large capacity and depth it may be built three pipes deep. The size of pipe generally used for these coils is ¾ in., as then the fittings do not take up an undue amount of room and the coil can be made to fit snugly to the walls of the fire box. Unless special provision has been made for the use of such a water heater by the maker of the stove the interstices of the coil and the space left at top and bottom should be filled with a good stove cement. This does not affect the heating qualities of the coil and prevents overheating of the plates of the range or leakage of smoke and gases to the room. As with the extension coil already spoken of the pitch to the outlet must be carefully watched. A dip at one of the bends caused by the action of screwing in a pipe or tightening a union is easily made and is often sufficient to retard the circulation to a serious extent.

A variation on the usual method of fitting the coil is shown in Fig. 14. Here the coil is shown suspended over the fire and in such a position as not to interfere with access to the fire box through the two stove lids. This is not likely to give as much satisfaction as the previous method, as the coil is subjected to radiant heat only and this is of much less value than direct contact with the glowing fuel. A coil of two pipes fitted to three sides of the fire box is shown in Fig. 15.

Proportioning Coils and Water Fronts.

The proportions of water fronts and coils are not subject to much variation to suit the needs of different households,

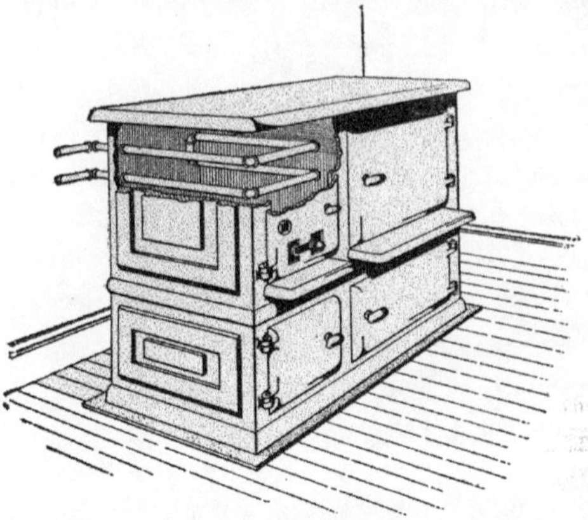

Fig. 15. A Coil on Three Sides of the Fire Box.

owing to the limitations which the construction of the stoves in which they are placed entail. It is, however, occasionally desirable to estimate the capacity of a water front or a coil before proceeding to install fixtures which would require a supply of hot water therefrom and which might entail an additional tax on the heating capacity which it was unable to carry. If the boiler be a forty gallon one and two hours be required in which to heat it to the desired temperature of say 110 deg. Fahr., it will be easy to compute the time required to heat any extra amount. As 1 B.t.u. is the equivalent of the heat required to raise the temperature of 1 lb. of water 1 deg. Fahr. when the water is at its point of maximum density—39 deg., it will be seen that to raise 40 gall., or 333 lbs. 70 deg. will require 23,310 B.t.u. If this is divided by 2 it is seen that the rate of heat transmission is 11,655 B.t.u. for the surface exposed. As the average size of the water front is about 5 in. by 15 in., that would mean a rate of transmission of 22,377 B.t.u. per sq. ft. per hour. This rate of transmission is high and will seldom be reached in a kitchen range water front unless the draft is good and a large fire is maintained. It is perfectly safe to estimate the proportions of a coil on a basis of 15,000 B.t.u. transmitted per sq. ft. per hour, as it is in more intimate contact with the hot fuel than a water front. It is only necessary to find the area of the pipe exposed to the fuel and calculate how many B.t.u. the coil will transmit from the fuel to the water, using a ratio of 15,000 B.t.u. per sq. ft. as a basis. When the number of heat units is found this should be divided by the number of degrees it is desired to raise the temperature of the water which will give the pounds that may be heated to that point and this again being divided by 8.3 the sum is found in gallons. To find the size of a coil or water front necessary to heat any stated quantity of water in a given time the process is reversed.

The consumption of fuel per sq. ft. of grate surface varies greatly according to the local conditions. In estimating the amount of fuel required to heat water in a kitchen range a consumption rate of 8 lb. per sq. ft. per hour should not be exceeded while in a water heater of special design the rate of combustion may not exceed 3 lbs. per sq. ft. per hour. A heat transmission of 8,000 B.t.u per pound of anthracite is generally estimated.

It is not advisable to estimate sizes on the maximum requirements, the average household only uses hot water at its maximum rate for very short periods and it is economical to cover these by an auxiliary heater as a rule. It must also be remembered that the whole contents of a boiler will not be at a uniform heat. The hottest water will be stored at the top of the boiler, while that at the bottom will be cold until the fire has been run for a considerable time. Therefore if water to a certain quantity is to be heated in a stipulated time the average temperature must be taken, as the hottest water will be drawn first.

The proportions of boilers for ordinary residences are not determined by any theoretical consideration but by what experience has shown to be correct. The ordinary residence with one bathroom and kitchen equipment is generally well served when a boiler of 40 gallons capacity is installed, and a 50-gallon boiler will serve a house with two or even three bathrooms very well.

In apartment and tenement houses served from a common storage tank it is common to provide 20 to 25 gallons per family up to 15 families; in a house accommodating 20 families, 20 to 22 gallons per family; for 25 families, 18 to 20 gallons per family, and over that 15 to 18 gallons is usually sufficient. If laundries are provided in the building the allowance should be increased about 30 per cent. to balance the increased demand for hot water on certain days.

CHAPTER IV.

Range Boiler Connections for Various Conditions.

While it is not possible to show an example of every possible method of connecting ranges it is imperative to show all of the boiler connections and methods of circulation that are

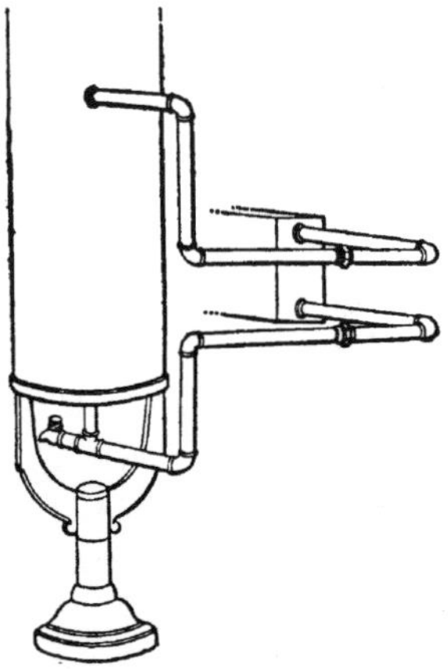

Fig. 16. Common Method of Connecting Range with Allowance for Expansion.

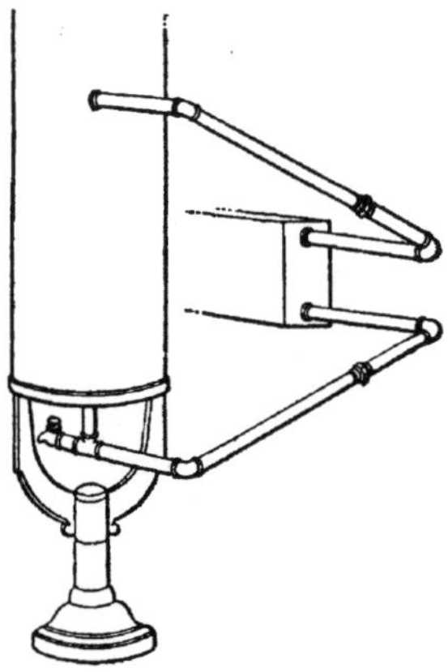

Fig. 17. A Method of Connecting a Range to Secure Quick Circulation.

likely to be met with in the work of installing hot water supply in large or small houses, hotels, and public buildings. There are many cases in which it is impossible to follow the stereotyped plan, the strict letter of the law as laid down in theoretical works on the subject. Where such departures are made in any of the examples here proposed the reason for it will be given, and where possible examples of installations which have been long enough under observation to afford definite proof of the claims set up for or against them will be illustrated.

For the first example is taken the simplest installation of all, the ordinary 30 or 40-gal. range boiler with regulation connection to water front, as shown in Fig. 16. Every plumber knows how to connect a range like this. He knows that carrying his pipes along in the manner illustrated will, as a rule, cause a circulation to take place. He knows that he is to allow for "swing" to prevent leaks at the joints through expansion, and that he is to provide a sediment cock or a tee with stop-cock and connection to the waste water system at the lowest point of the boiler connections.

But every plumber does not know what to do when the circulation refuses to materialize; or when the boiler is hot at

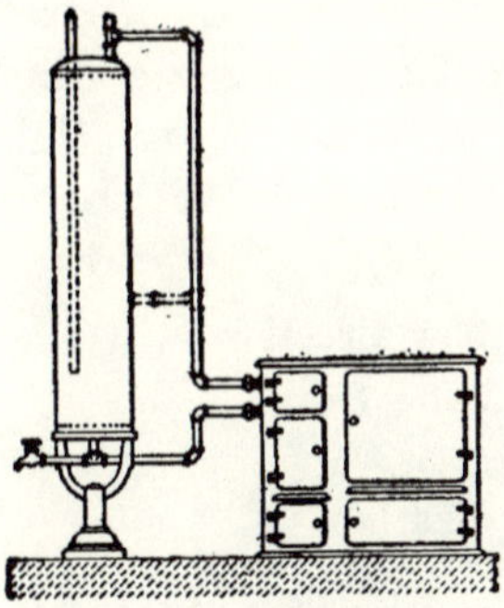

Fig. 18. A Quick Heating Connection.

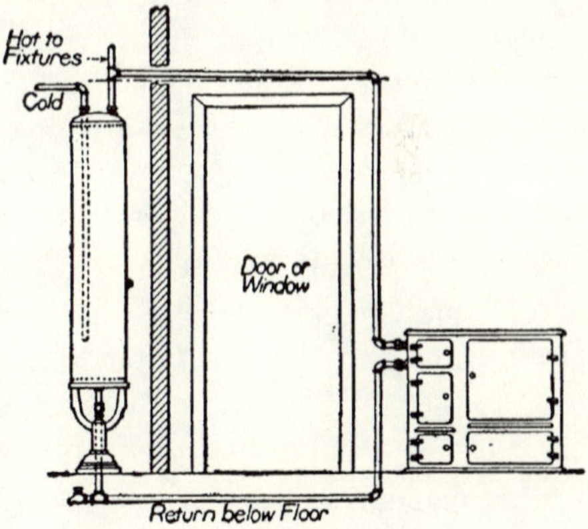

Fig. 19. Connection to Clear a Door or Window.

times and cold under apparently the same conditions. Also what to do when heating of the boiler is accompanied by pounding and rattling noises and vibration of the boiler.

Pounding.

The man who is up in the theory of his business will promptly begin to analyze the symptoms. He will go carefully over them and eliminate "possibles" one by one until he finds the cause of the trouble. No waste of time in experimenting at haphazard; just a methodical application of the knowledge he has gained from his books and papers and from his own deductions and observations in his daily work.

Pounding is caused in several different ways. By overheating, owing to the water front or coil being of larger size than is necessary for the size of boiler used. There is no cure for this except increasing storage capacity or reducing heating surface by inserting a brick or other heat-resisting medium between a portion of the water front or coil and the fire. A radiator is sometimes fitted where such can be used with advantage. This provides a ready means of collecting and dissipating satisfactorily the excess heat.

Pounding may also be caused by defective circulation through sediment collecting in the water front, in circulating pipes or in the boiler. This prevents a proper and continuous supply of cold water to the water front, and the result is that the water is overheated, steam bubbles form, and on encountering the cooler water in the boiler in their circulating path, suddenly condense, and partial vacuums are formed with the attendant rumbling and snapping noises that are produced by the water rushing in to fill them.

Insufficient pitch, or pitch in the wrong direction, will also produce the same trouble through collection of air.

Connection to Facilitate Heating.

Failure to properly heat the boiler is occasionally difficult to account for. In the first place it may be poor firing, which will never be admitted by the complainant; it may be poor coal, poor draft in the chimney, or insufficient heating surface.

When trouble along these lines is encountered it is often advisable to connect up the boiler after the manner shown in Fig. 17. This, although a "stiff" connection, is quite allowable. In fact, in some respects it has the advantage of the other method, as there is a minimum of frictional resistance to the circulation and the natural flow or rise of the heated particles of water is assisted by the sharper pitch of the pipes.

The author has had occasion to enlarge the size also, and a 1-in. pipe instead of the customary ¾ in. will make quite a considerable difference in the efficiency of the system.

Then the size of the smoke pipe must be considered and the draft tested to decide if the chimney is faulty.

The author has also found a partial stoppage in a water

front, caused by parts of the core used in casting being left in it, that was impeding circulation greatly, but yet not enough to cause pounding or snapping through overheating.

It is always wise to examine thoroughly a new water front before installing and to pass a bent wire through to prove the passage at the end is clear. Also, in connecting to an old boiler,

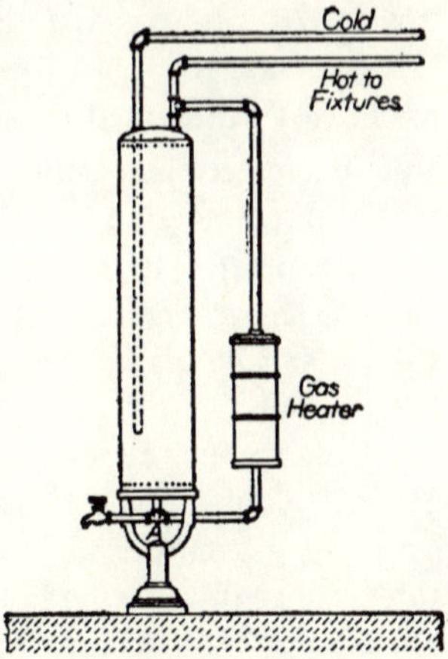

Fig. 20. Connection for a Gas Water Heater.

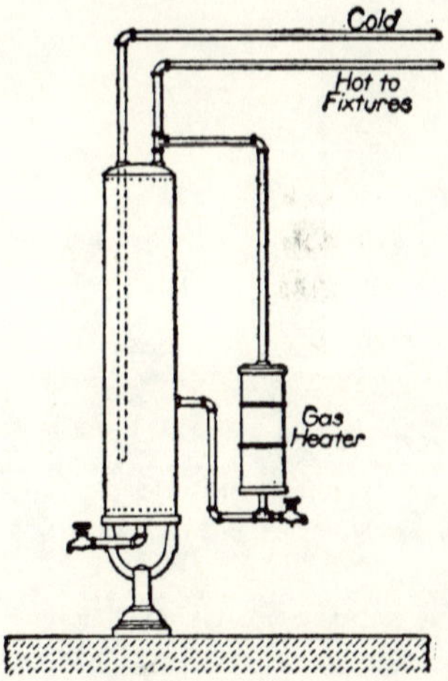

Fig. 21. Connection of Heater to Avoid Sediment in Boiler.

see that there is no deposit of ooze or rust on the bottom, as this will often be dense enough to prevent the return circulation to the water front. As much as 18 in. has been found in an old boiler in investigating the cause of non-heating. Also make sure that the supply tube is in good condition and clear.

Quick Heating Connections.

Variations on the regular methods of connecting an ordinary range boiler may be made to suit special conditions. In fact, some conditions demand a departure from the ordinary methods. Such a condition may arise where the water front or whatever the heating medium may be is so high that a proper pitch cannot be given to the flow pipe and allow of its being connected to the side opening in the boiler. In this case the side connection is plugged and the flow pipe carried to a point immediately above the boiler, where it is connected to a tee on the house supply as it leaves the boiler.

A feature of this style of connection is that the hot water being delivered to the upper part of the boiler is more quickly available. As it does not have to pass through the colder water in the boiler, transmitting heat by convection to the contents, it follows that a larger quantity of really hot water is also available in shorter time than with the other connection although no gain would be shown in heating the *whole contents* of the boiler to a given temperature.

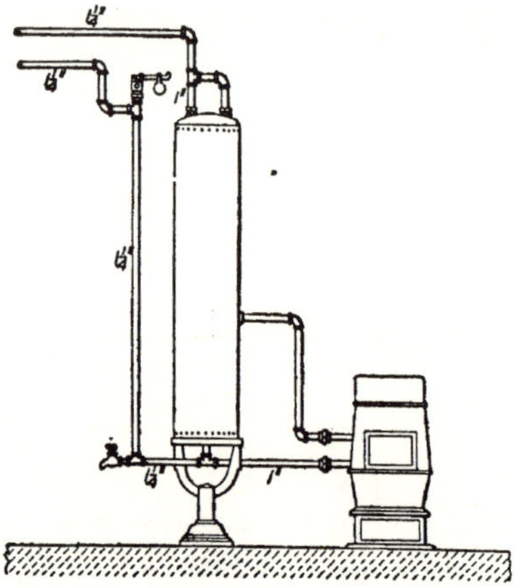

Fig. 22. Connection to Afford Free Flow Under Low Pressure.

This style of connection is generally known as a "quick-heating connection," and where small quantities of very hot water are required at frequent intervals is a very satisfactory one to adopt. A combination of the good features of both may be obtained by connecting with the side opening also, as shown by the dotted line in Fig. 18. This, while allowing of the hottest water being stored at the top of the boiler for instant use, permits a proportion of the flow from the water front to pass into the main body and, by mixing with the colder water to bring the temperature of the whole to a more or less equal state in a shorter time when no water is being drawn.

Connection for Boiler With Door Intervening.

In many instances where the boiler has to be placed in a closet or other apartment, a long run of pipe is necessary, and often there is some obstruction which will not permit of a connection to the side opening, even if pitch could be got for the long run. In this case also the connection would be made to the top of the boiler, plugging the side outlet. Where a door or window has to be crossed the same method is advocated.

When a pipe is carried up to clear an obstacle of this

nature it should enter the house supply over the boiler at a slightly higher level than the ell at which it makes the change from the perpendicular. If it pitches downward from the ell air will lodge at the high point, and if pressure is low will probably cause an air lock. The return pipe may be dropped below the boiler level if necessary to clear a door or other obstruction, the sediment cock, of course, being placed at the lowest

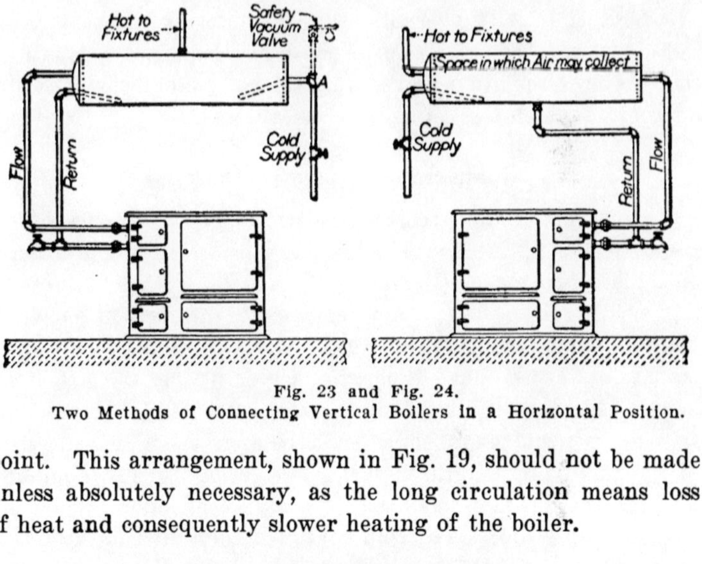

Fig. 23 and Fig. 24.
Two Methods of Connecting Vertical Boilers in a Horizontal Position.

point. This arrangement, shown in Fig. 19, should not be made unless absolutely necessary, as the long circulation means loss of heat and consequently slower heating of the boiler.

Connection for Gas Heaters.

A more or less common means of heating kitchen boilers is that by a small gas heater, consisting of a coil of pipe, or a series of hollow metal disks through which water is passed, and a bunsen burner, the whole being enclosed in a sheet metal or cast iron cylinder. This appliance is fitted directly to the side of the boiler, as shown in Figs. 20 and 21, and has variations in the way of automatic gas control and other features peculiar to the several makers' designs.

The most commonly used connection is that shown in Fig. 20. This admits of the whole contents of the boiler being heated, but is open to objection on the score that sediment from the boiler may be carried along and deposited in the coils or disks of the heater, leading to a stoppage of circulation and overheating of water with attendant pounding, and possibly to a bursting of the disks or coil. This objection may be overcome to a great extent by fitting a sediment chamber at the tee marked A. If this is given proper attention and the deposit periodically removed no trouble will be experienced.

A plan which is often recommended is to take the return circulation from the side opening of the boiler, as shown in Fig. 21. This effectually prevents sediment entering the heater, but reduces the storage capacity of the boiler by about one-half, as there is no circulation and storage of hot water below the side opening to which the return pipe is connected.

Connection for Low Pressures.

Occasionally the problem of providing a rapid flow of hot water at the fixtures with a low head of water is encountered and there are various ways and means of solving it more or less successfully. The author has found the method shown in Fig. 22 entirely satisfactory as it does not call for anything outside of the standard type of boiler or fittings.

It shows a regular tank heater connected to a standard tapped vertical range boiler in a manner calculated to allow of the maximum flow through it. The supply tube is eliminated and the two top tappings used for a twin connection to a 1¼-in. line of supply pipe. The cold supply is connected to the return connection to the boiler and is also 1¼ in. diameter until it is reduced at the bottom tee to 1 in. to suit the standard tapping. The minimum frictional resistance to the flow is thus encountered as the combined area of the outlets more than equals the area of the supply line and the same is true of the cold supply and the inlets to the boiler.

A combined safety and vacuum valve fitted as shown entirely prevents siphonage should the supply be intermittent and is also a useful adjunct through its being calculated to open under an excessive pressure or under a sudden shock from a quick closing valve. Much has been written concerning the undesirability of connecting the cold supply to the boiler in this fashion, the statement being made that the cold water entering is liable to flow clear to the top of the boiler and cool off the water leaving it. This may be true under a heavy pressure, but there would be no advantage in using this connection in that case, and under light or moderate pressures the author has found this method satisfactory in every way.

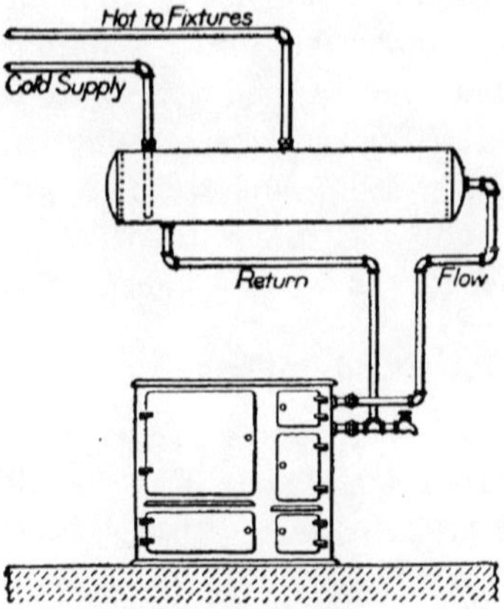

Fig. 25. Common Method of Connecting Horizontal Boilers.

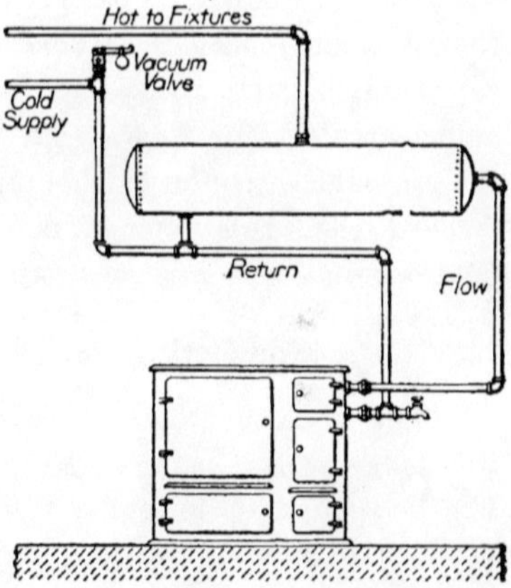

Fig. 26. A Method that Requires Only Three Tappings

Overhead Horizontal Boilers.

Lack of room or some similar consideration often favors the placing of a range boiler in a horizontal position overhead. If the standard vertical type of boiler is placed in an overhead horizontal position the result will not be entirely satisfactory.

The best method that can be adopted in this case is open to objection and the only course is to choose the least of the evils. That which has the least against it is shown in Fig. 23 and a glance at the sketch will show at once that it is not free from fault. The circulating pipes from the water front enter the boiler through the top tappings, the return having a dip pipe to ensure the whole contents of the boiler being circulated. The supply should enter the bottom tapping and also, of course, have a dip pipe, with the usual vent hole drilled in it.

A decided improvement over the vent hole is a combination vacuum and safety valve fitted as shown by dotted lines in Fig. 23, as half the contents of the boiler could be siphoned out before the vent hole would break the vacuum. It will be noticed that the boiler cannot be properly washed out either.

While the dip pipe on the return connection will siphon out the bulk of the contents it does not remove sludge and other deposit and this cannot well be scoured out in the ordinary manner.

The other method of connecting, however, shown in Fig. 24, while it does not possess this objection, has the graver one that a steam pocket can be formed above the outlet, and thus cause hammering and other serious troubles, and this fault is sufficiently serious to warrant the recommendation of adopting the alternative connection before described. A better plan still is to discard this type of boiler and procure one tapped in a manner that will insure a satisfactory job when connected up.

Connecting Regular Horizontal Boilers.

The usual method of connecting a horizontal boiler with the range water back, or heater, is that shown in Fig. 25. It will be seen that the boiler differs from those used in a vertical position in having the tappings for cold supply and return in the sides instead of the ends. This permits of all the contents being drained when it is necessary to wash out the boiler, and also insures circulation of the entire body of water.

The cold supply, entering through the top, is fitted with a supply tube in the regular manner, and a hole drilled in it a few inches from the boiler union will effectually prevent siphonage should the supply fail in the city mains or be cut off by a stopcock with waste outlet at a lower level than the boiler.

An alternative method of connecting the boiler, using only three tappings, is shown in Fig. 26. This is a good system to adopt where the pressure is low, but where a very heavy pressure is carried in the cold supply the water entering the boiler through the return connection may be forced to the upper part of the boiler and cool off the hot water stored there. A vacuum

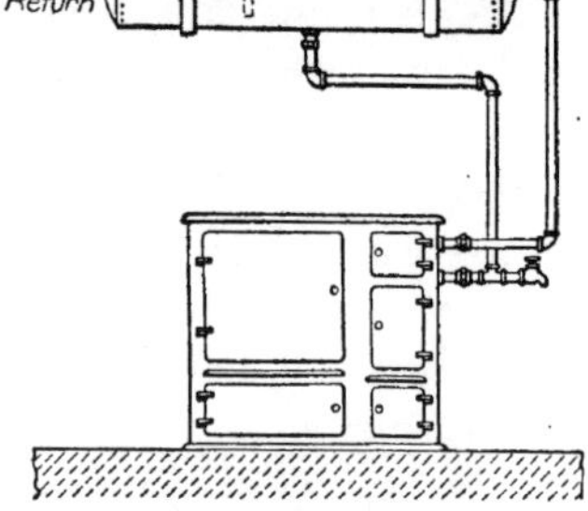

Fig. 27. Method Designed to Favor Quick Supply to Fixtures.

valve is used in connecting in this manner the cold supply from the street main. This will open immediately the pressure falls below that obtaining in the boiler and thus effectually prevent siphonage back into the mains. This valve must, of course, be fitted on a level above the top of boiler. Should the supply come from an overhead tank the valve is unnecessary unless a faucet or stopcock with waste outlet is fitted on the same line at a lower level than the boiler.

A little variation on the method of connecting a horizontal boiler is shown in Fig. 27. It will be noted that one tube is connected with the cold water supply to the boiler and runs down inside of the boiler to a point near the bottom. The other opening at the top of the boiler is for the hot-water service connection. The tube at one end of the boiler is designed to bring the hot water from the water back near the top of the boiler, so that it will find its way immediately to the hot-water service piping. The opening in the other end is frequently used for a return or circulating pipe, so as to keep the hot water moving, so that as soon as the hot-water faucet is opened at any fixture hot water will flow without any waste of the water, as is the case when the cold water in the pipes has to be drawn off to allow the hot water to reach the faucet, as is necessary where there is no circulating pipe. The bottom opening in the boiler naturally connects with the bottom or return opening in the water back in the range.

Example of a Steam Heated Boiler Which Was Not Satisfactory.

While on the subject of horizontal boilers we may touch on those of this type heated by steam coils. In very many cases these are placed in the basement close to the steam boiler, and on a level very slightly above the water line in the boiler. If this happens to be of the low pressure domestic heating type, care must be taken to connect up the coil so that condensation will not be held up in it and so lower its efficiency.

In Fig. 28 is shown the style of connection referred to and which illustrates a case the author has in mind. This coil had never given satisfaction. The heater is in the basement of the gymnasium of a large preparatory school and steam is not maintained in the boiler continuously. A battery of twenty-five showers is supplied from the hot-water boiler, and when these are in use an extra large quantity of hot water is called for. The pressure of steam necessary to heat the building is only about 1 to 2 lb. and the complaint was that at that pressure the coil was ineffective, the return being cold to the touch and the heating effect on the boiler inappreciable.

There was a check valve on a horizontal part of the steam connection just above water level of boiler, and the author concluded this had stuck. Finding it in good condition, he con-

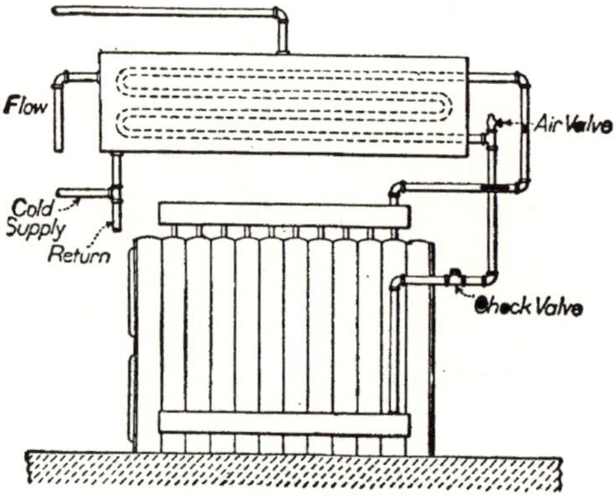

Fig. 28. A Boiler Fitted Very Close to Water Line of Steam Boiler.

cluded that the rapid condensation in the coil lowered the pressure at the return end so much (at low boiler pressures) that the water of condensation did not have head enough to overcome the check before the coil stood partly full. The author prepared to fit a small equalizing pipe to counteract this, and in doing so lowered the check below the water line of the heater. On plugging the tees and trying out he found that the coil worked perfectly without it, the extra drop evidently being sufficient to open the valve. He would have removed the check entirely had the boiler been in continuous use or its level higher above the water line of the steam heater. A Breckenridge air valve, which was found in good condition, was carefully set so as to insure quick egress of air when steam was raised and the results are now satisfactory.

This example, while not very common, may be met with from time to time, and the liability to failure through placing ler at too low a level must not be overlooked. Where .ce available prevents a safe difference of levels being special care must be given to proportioning the steam and returns so that the pressure at the supply end of will be as nearly equal to that at the return end as . The equalizing pipe mentioned is a pipe of small r connected to the flow and return connections before ply enters the coil in the tank for the purpose of se- that the pressure will be equal in each, no matter what ge pressure may be. This effectually prevents water up into the coils from the steam boiler when the water torage tank is cold and the steam is being condensed as as it is supplied.

CHAPTER V.

Variations in Connections to Suit Special Requirements.

The problem of making hot water flow downward from a range to a boiler on the floor below it may well be called the plumber's fifth proposition—the "Pons Asinorum," for it has puzzled more young mechanics, perhaps old mechanics also, than almost any other that we have to solve. It has been explained so often that it would seem as if every member of the trade would be familiar with it and yet an inquiry turns up with unfailing regularity in the trade papers every few weeks. Fig. 29 shows the usual method of connecting a boiler in this position when the supply is direct from a city pressure main. The flow pipe from water front is carried up to a point at or above the ceiling of the kitchen and then turned down to connect into the top tapping of the boiler, an air cock being fitted at the highest point to draw off any air that may accumulate in the loop or a fixture above this level supplied from a branch taken from the top. The branch to supply fixtures is usually taken from a point a little above the boiler connection as shown in the illustration in Fig. 29.

Circulation is established and maintained between the boiler and range water front by the cooling effect that the loop in the piping affords. Here are two columns of water of equal height. That in the descending leg, starting from the top of the loop, is obviously the cooler and therefore the more dense and by the law of gravitation must fall to balance the other and less dense column of which the water front forms a part. Thus the hot water is pushed up and around the loop and, becoming slightly cooler as it progresses, descends into the boiler.

The results obtained in this system are not generally so satisfactory as with the boiler on the same or a higher level than the heater, owing to the slower circulation. A safe rule for the height the loop should be made reads, "Twice the height above the water front that the boiler is below it." Any extra height over this will tend to increase the efficiency and hasten the circulation.

A point that is sometimes lost sight of in installing a boiler in this way is the danger of emptying the water front by siph-

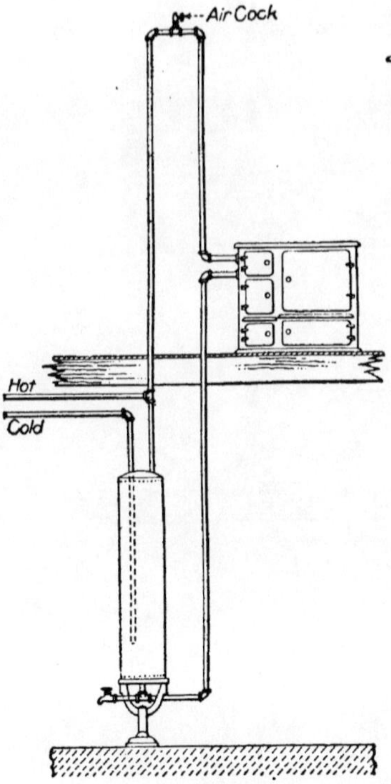

Fig. 29. Common Method of Connecting Boiler Below Stove.

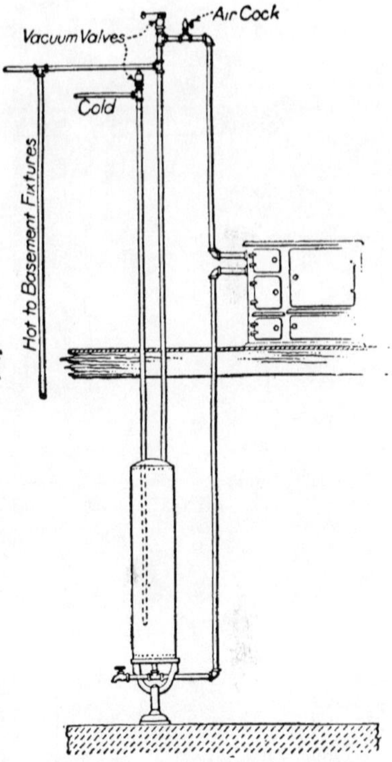

Fig. 30. Connection to Prevent Drawing Water Below Level of Stove.

onage should the pressure in the city mains fail. Just as soon as the pressure goes down the water must fall back into the boiler through the return connection leaving the water front empty, and if a very hot fire is maintained trouble will assuredly follow. Also, if there is a branch from the hot-water pipe to a fixture in the basement it is possible to draw the water back in the same manner should the main house valve be closed at any time. To get over this the connection can be made as shown in Fig. 30. Here both the hot and cold-supply lines are shown dropping from a point above the level of the water front, the branch to the basement fixtures being also taken from this level. Vacuum valves fitted on both lines at the highest vertical point will effectually prevent siphonage of the boiler back through the cold supply from the city main pipes or through the basement hot-

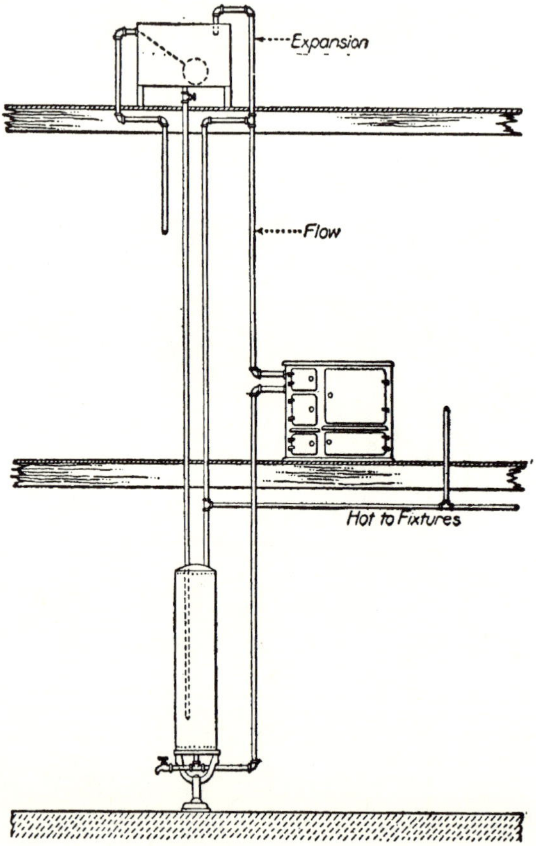

Fig. 31. Boiler on Floor Below Stove Supplied From Attic Tank.

water faucet. This is an important point and should always have consideration, as although the system shown in Fig. 29 will work perfectly in providing hot water the danger from the above-mentioned causes is always present.

Fig. 31 shows the usual method of connecting a boiler below the level of a heater when a tank supply is used. There is nothing differing greatly from the method used in connecting when the supply is from the city mains. Instead of using an air cock or connecting a branch to some fixture to the highest point of the loop to relieve it of air, an expansion pipe is connected here

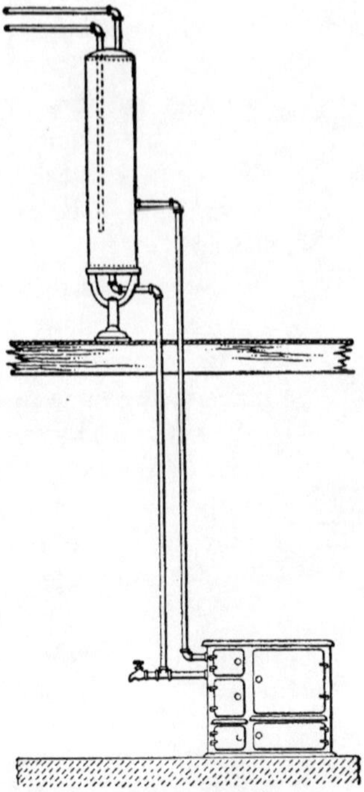

Fig. 32. Connection for Boiler on Floor Above Stove.

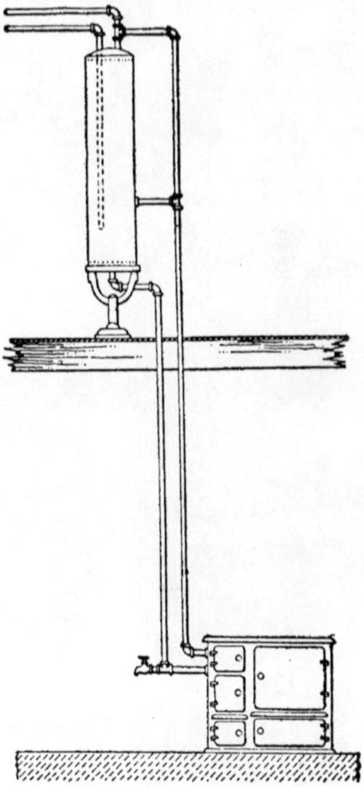

Fig. 33. A Method of Connecting to Favor Quick Heating.

and carried up and turned over the top of the tank. This allows air to escape as it is formed and also serves as a ready means of escape for the water when expanded by heat should the supply valve happen to be closed. If the tank supply should fail the same danger of emptying the water front would exist if a branch be taken to a basement fixture from the boiler and therefore the connection should be made at a point above the level of the water front, or carried up to a point above it and provision made to guard against siphonage before dropping to the fixture in the basement.

Connecting Boilers on Floor Above Heater.

The problem shown in Figs. 32 and 33 scarcely deserves to be classed as a problem as there is barely room for any error in making such connections. Occasionally a request for the proper method of connecting a boiler on the floor above the heater is made and as the examples shown are designed to cover all combinations in the practice of the average mechanic it cannot well be overlooked.

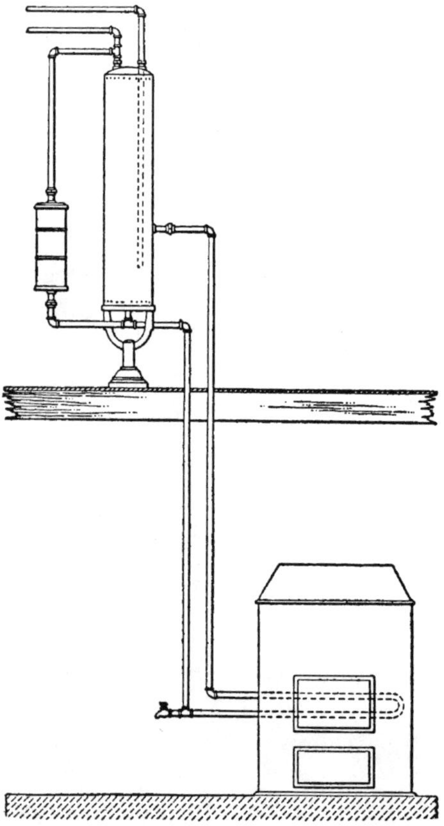

Fig. 34. Connections for Furnace Coil and Gas Heater.

As will be seen in Fig. 32, the flow and return connections are made in the same manner as if the range were on the same floor as the boiler, but the pipes are continued down through the floor to the water front or coil. The sediment cock is placed at the lowest point of the return piping, that is, where the turn is made to connect to the water front or heater. The supplies to the various fixtures must be taken from the top of boiler in the usual manner. Supplying a fixture on the floor below the boiler direct from the circulating pipes must never be permitted as there would be a liability to empty the boiler without the knowledge of the user should the cold supply for any reason be cut off. This would entail danger of bursting the water front when the cold water is turned in again should a strong fire have overheated and evaporated the water left below the level of the branch to the fixture.

Fig. 33 shows a quick heating connection made under the same conditions. This is a highly efficient method to adopt if a good supply of very hot water is required as the height of the circulating system induces a rapid movement and the flow to the top of boiler stores the hottest water there. The side connection tapping may be plugged or used to make the combination connection as shown. It is advisable to use it to secure the best results.

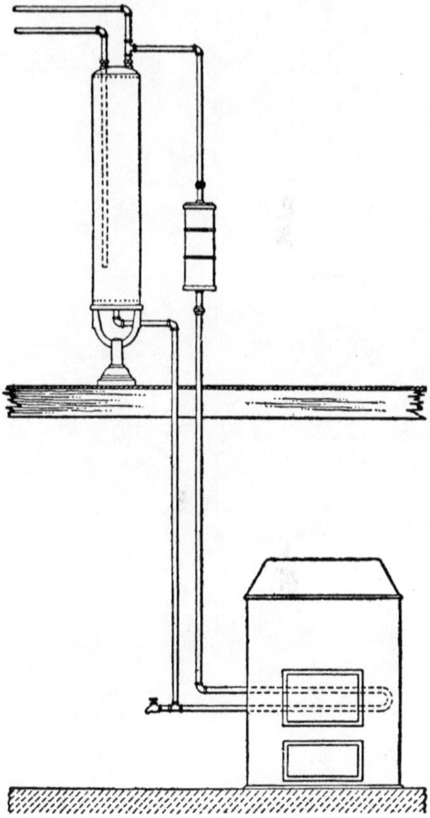

Fig. 35. Connections Giving Continuous Flow Through Gas Heater.

Connections for Coil and Gas Heater.

The most common application of placing the boiler on the floor above the heater is where an independent hot-water heater, a coil in a heating furnace or a cook stove in a summer kitchen in the basement is the heating appliance provided. When such is the case a gas heater is generally used in addition. This provides a ready means of heating the boiler when the heating furnace is not required or when it is not desirable to maintain fire in the cook stove at all times of the day.

These gas heaters are usually connected as shown in Fig. 34, the side connection to the boiler being used for the flow from the coil or water front in the basement, while that from the gas heater enters the pipe leaving the boiler to supply hot water to the various fixtures. An alternative method of connecting is shown in Fig. 35. This greatly simplifies the piping and is generally efficient. The coils or discs of the gas heater act as a radiator which cools off to some extent the water flowing from the furnace coil. The loss is partly compensated by the slightly increased rate of circulation obtained by the increased height of the flow pipe due to its entering the top tapping of the boiler and may in most instances be disregarded as the casing of the gas heater prevents the radiating surfaces of the coils or discs from transmitting as much heat to the air as they would do if fully exposed to it. There is in reality little to differentiate one method from the other and convenience alone will dictate which is the best to adopt for any particular set of conditions.

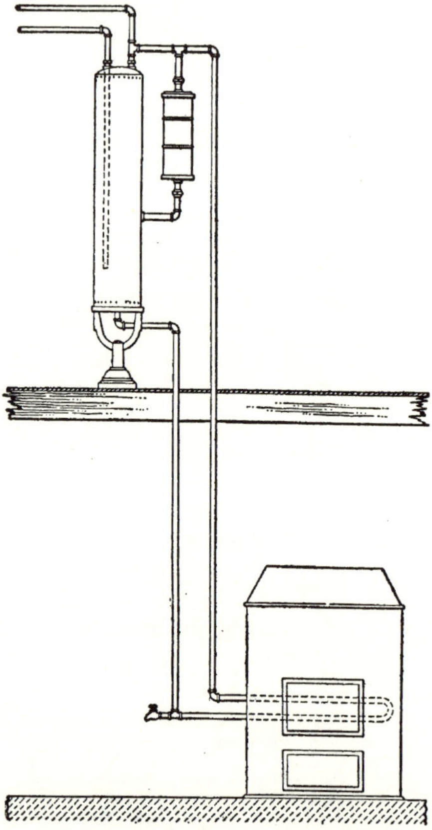

Fig. 36. Connection for Gas Heater to Reduce Storage and Raise Temperature of Water.

Still another method is shown in Fig. 36. The gas heater in this instance has its return connection made to the side tapping of boiler. This obviates any chance of sediment finding its way into the gas heater and also reduces the amount of water stored by the heater so that the contents are at a higher temperature in the upper part of the boiler where it is ready to be drawn as soon as a faucet is opened.

Adding Storage Capacity.

To add to the capacity of a hot water system it is often convenient to install an extra boiler. This provides additional storage capacity and if the water front is of sufficient size this will often be found a satisfactory arrangement. The manner in which a horizontal boiler may be connected to one already

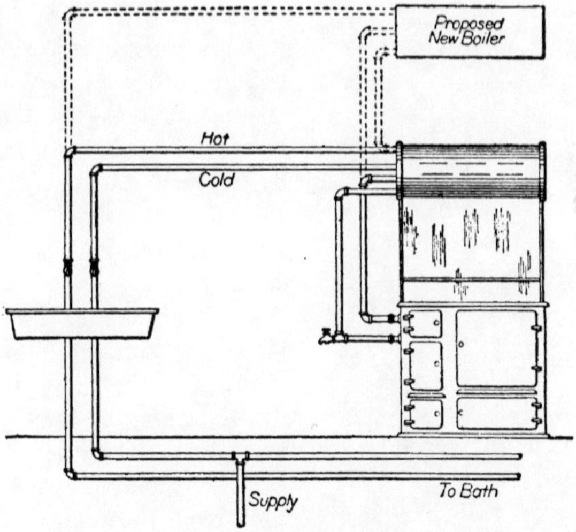

Fig. 37. Method of Connecting Additional Boiler.

in position is shown in Fig. 37. The new piping is indicated by the dotted lines. The present hot water service pipe should be disconnected from the lower boiler and connected at the top of the upper boiler. A pipe should be carried from the top opening in the lower boiler to the bottom opening in the upper boiler, and this pipe should be of the full size of the openings in the boilers to ensure free circulation.

The pipe bringing the hot water from the water back should be disconnected from the lower boiler and the opening stopped with a plug. This hot water pipe should then be connected with the middle opening of the upper boiler. This method of piping will allow the cold water in the upper boiler to pass to the lower boiler as the hot water enters it from the water back, and the cold water will pass on to the water back and a circulation be kept up. There should be no trouble from the use of two boilers piped in this way, providing the water back has the heating capacity to heat the extra amount of water contained in the upper boiler.

A vertical boiler may be connected as shown in Fig. 38, when the circumstances are similar, that is, when the water back is sufficiently large to take care of the additional tax on it. So that the flow from each will be equalized the connections are made from the top of the boiler into a tee about half way between the boilers. The cold water supply is led into the two boilers in the same manner. This ensures that the friction in the supply lines to and from each boiler will be the same and therefore the draft on each is likely to be equal. If it is desired to favor one more than the other such as for instance when one boiler is much larger than the other and the water drawn from the smaller would be cold before the other was emptied of all the hot water it contained, the connections could be such, by suitable sizing of the pipes, that this could be done automatically. Valves may also be inserted and set to pass only the quantity desired from each.

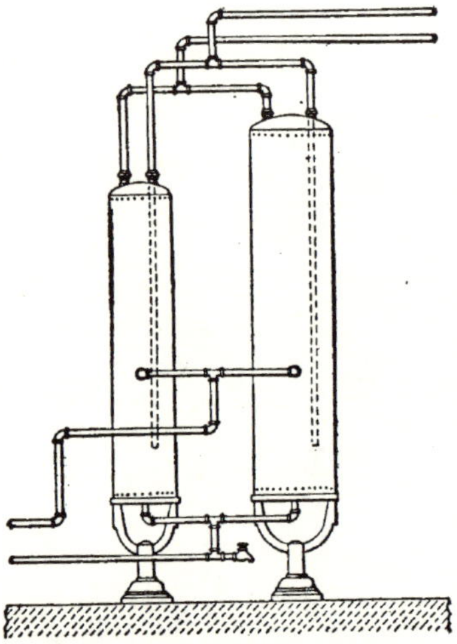

Fig. 38. Connections Made to Equalize Flow from Additional Boiler.

When an increase in storage capacity of hot water is desired it is generally more satisfactory in the end to install a larger boiler rather than an auxiliary to the existing one, as there is then no difficulty in securing positive circulation from the water front. But there are many cases when to do so is not feasible, owing to local features of construction or higher cost. When an auxiliary boiler to the existing one is decided upon local conditions again will dictate the best method of making the connections. An important consideration to bear in mind when arranging the piping for a connection like this is the desirability of equalizing the flow to each boiler. Means should be taken, by the use of the proper fittings and proportioning of pipes, to insure as nearly as possible the same conditions in each water front, for if the resistance to the flow by friction, etc., is about the same in each branch of the flow pipe there will be a better chance of equal heating of the boilers.

Two Boilers Connected to One Water Back.

Figs. 39 and 40 illustrate the two most common methods of connecting up two boilers to one water back on the same floor level. When the boilers are close together there is nothing very hard to overcome in securing a satisfactory circulation to each, and in this case the connection best suited is that shown in Fig. 39. A distributing T or a Y can be fitted at the branch to the first boiler to promote equal distribution, but the reduction from 1 in. to ¾ in. at this point tends to secure this satisfactorily. Good pitch should be given to these pipes, crooked threads being cut to keep the vertical parts plumb, and the connection to the side tappings of the boiler, if the tapping is 1 in., should be made by a reducing ell rather than by a bushing. This is done to eliminate friction as far as possible.

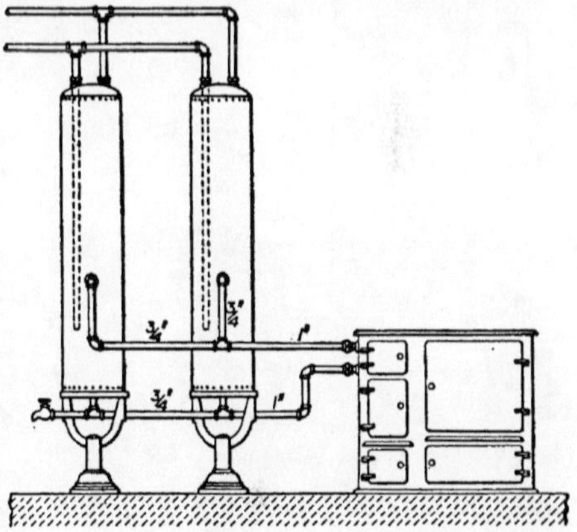

Fig. 39. Connection for Two Boilers to One Water Back.

Should the boilers be set far apart it may be impossible to secure sufficient pitch to connect to the side tappings of both boilers. In this case, one of them, or both of them if desired, can be connected as in the case of "a quick-heating connection." Fig. 40 shows two boilers with one of them connected in this manner. A distributing tee can be used at the connection to this boiler if it is thought necessary, but as a rule the extra height of the circulation compensates for any tendency to restricted circulation by reason of the velocity of the flow past the tee retarding the flow to this boiler. Either of the boilers may, of course, have the side connection, but generally that nearest the range will have the side connection and that farthest away the quick-heating connection. The judgment of the fitter must be used to determine which it will be, but always the equalizing of the flow ought to be kept in mind.

Fig. 41 shows two boilers on different floors connected to one heater. This is not very commonly done, but where a powerful heater is used it is quite successful and often the construction of the building is such as to prohibit any other method.

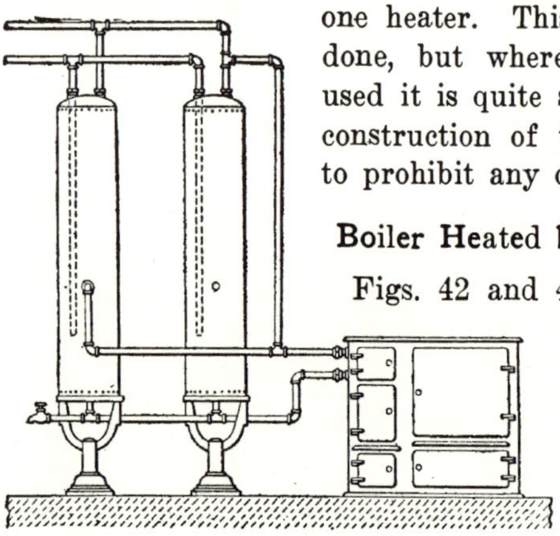

Fig. 40. Connection to Equalize Flow to Boilers.

Boiler Heated by Two Water Backs.

Figs. 42 and 43 show exactly the reverse conditions, these being examples of the methods of connecting two heaters to one boiler. The most commonly met with combination is where a laundry stove and kitchen range are connected to one boiler, both heaters being set close to the boiler. Again, where much cooking has to be done two ranges are commonly used with the water backs connected up as shown in Figs. 42 and 43.

It goes without saying that it is not good practice to connect more than one heater, if they are to be in more or less constant use, to the ordinary standard sized kitchen boiler, as overheating will be sure to occur. If the boiler is of 60-gal. capacity, or over, such trouble need not be apprehended. Fig. 42 shows one stove connected in the usual way to side connection of boiler and the other through the top tapping. It is always best, if circumstances will permit, to connect the range farthest away from the boiler through the top connection, as the quicker flow in the pipe counterbalances in some degree the longer path it has to take and so better equalizes the work each water front has to do.

When the method shown in Fig. 43 is used great care should be taken to get sufficient pitch to the pipes. There is a tendency toward retarding the flow at the point marked A, and this will often cause rumbling and snapping sounds in one or other of the water fronts, through the water becoming overheated.

By increasing the size of the pipe at the tee this may be largely avoided and the use of a Y as shown still further improves the connection. It is never, for obvious reasons, advisable to fit valves on the lines between the water fronts and the boiler. If one of the stoves is in a room which will be closed at certain seasons of the year, as is common in some localities, it is better to fit a second boiler which will be heated from the stove in that room than to connect it with the boiler in the all-the-year-round kitchen. This will permit the water to be drawn off the water front and its connections and the closing up of the summer kitchen entirely if desired. The only valves that will be required will be on the cold and hot supplies to the boiler in the summer kitchen and the fitting of these need add no risks to the operation of the stoves as would be the case if valves were fitted on the connections and both stoves connected with one boiler.

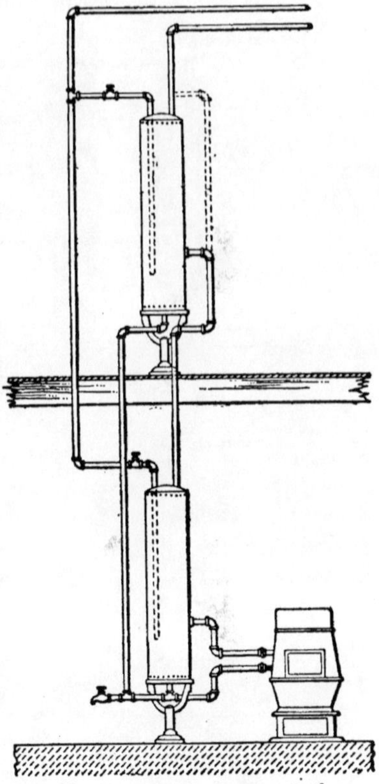

Fig. 41. Connections for Two Boilers on Different Levels.

Triple Connection to One Boiler.

A triple connection consisting of two ranges on the first floor with a tank heater in the basement is shown in Fig. 44. The utmost care is required in connecting the upper ranges so that the long flow and return pipes will pitch correctly. As shown, the connections of the tank heater provide for a continuous flow through the water front to the boiler while the second range on the first floor connects to the top of the boiler. An alternative method is shown in the smaller sketch. Here the three flow pipes are brought together and enter the side inlet of the boiler. The size of the pipes should be proportioned so that the flow

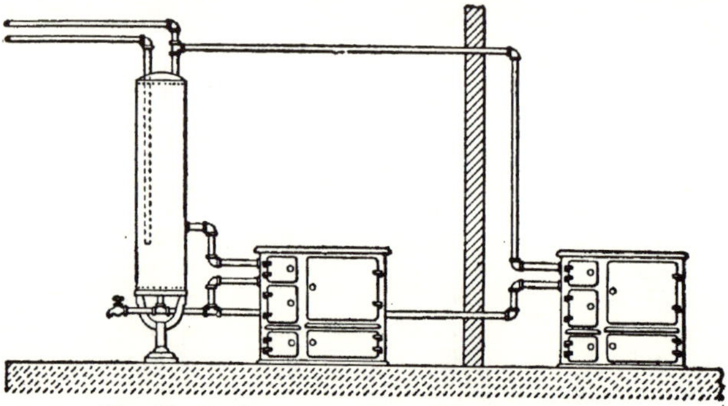

Fig. 42. Boiler Heated from Two Water Backs.

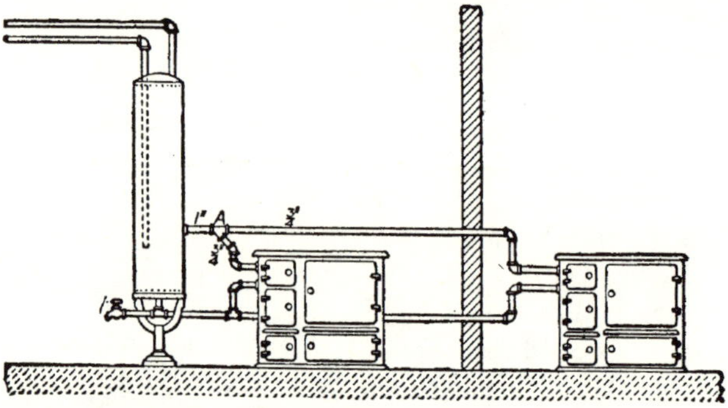

Fig. 43. Another Method of Making the Connection.

from one heater will not retard that from the other. This would lead to overheating and snapping sounds would be evident.

As in the case of the connections of two stoves on the same level to one boiler it is better to use two boilers than to attempt to cut out one of the three water fronts at such times as it might be desired to discontinue its use. The provision of valves on the connections may lead to accident through neglect and always obstructs the circulation to some extent. If the extra boilers are set in a summer kitchen the supply pipes may be run in a position to facilitate draining or protection from frost, and valves may be placed on the supply lines with less chance of being overlooked. A special fitting may be used at the connection to the top of the kitchen boiler should the circulation

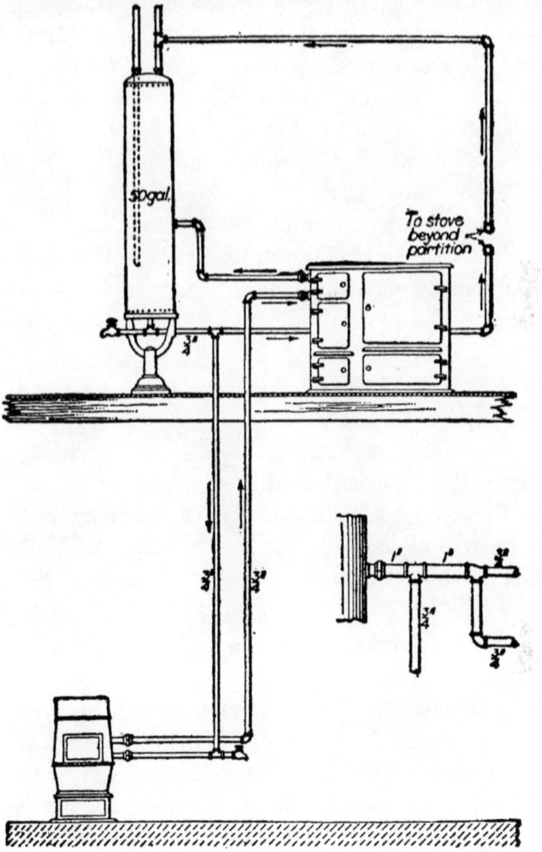

Fig. 44. Connecting Boiler from Three Water Backs.

be brought to there from the farthest away water front. This is designed in such a manner as to admit the hot water to the upper part of the boiler while preventing it from being drawn away without entering the boiler when a faucet is opened on the supply line. There is some chance of short circuiting the boiler and of drawing cold water directly through the water front and circulating pipe in some cases and this connection is designed to avoid such being possible.

CHAPTER VI.

Multiple Connections with Tank and Pressure Supply.

In any system of hot-water supply using one boiler and two heaters a more positive and reliable circulation will be obtained if one of the heaters is on the floor below the boiler, as it is easier to install the piping so that the circulation from each water front is assisted rather than retarded by the other. Fig. 45 shows such a system. The circulation through the two water fronts is continuous and either of the stoves or both may be used at one time without in any way changing the circulating path of the hot water. The piping is neat and is easily installed, and this is probably the most suitable method of making the connections for ordinary requirements. A technical objection may be offered to passing water through a front that may be cold but the cooling effect may be ignored as it is hardly appreciable.

Where a different style of connection is desired that shown in Fig. 46 will be found satisfactory. The circulation from each heater is separate and distinct and will not interfere with each other. It will be noticed that the upper stove is connected through the top tapping of the boiler, so that the circulation in each case will have about the same speed. Many fitters prefer this even though it entails more work and material than the style shown in Fig. 45.

Still another method may be adopted. The return pipes are fitted in the same manner as in Fig. 46, but the flow pipes are connected and enter the boiler through the side tapping, as shown in the small sketch in Fig. 44. When this is done the pipe should be enlarged at the junction and connected with the boiler at the largest size the tapping will take. If this is done the retardation of the circulation from either one of the stoves by the other will not be so likely to take place. There are many cases where it is absolutely necessary to use the side tapping in this manner and this care to enlarge the pipe will make all the difference between a satisfactory job and a failure.

When the boiler is on the lowest floor, one of the heaters being on the same level and one of them on the floor above, the circumstances are somewhat different. We have already shown the correct method of connecting up a boiler from a water front above it and all that is necessary to add another heater is to use the side tapping for the connection from the lower one. Fig. 47 shows how this is usually done. Entire satisfaction will result, at least the job could not be bettered, when the supply to the fixtures is taken direct from the flow pipe at the top of the boiler. But when the supplies to the fixtures are taken from a secondary circulating system, and the flow to this secondary loop is taken from this point, there is a liability to failure. At this point marked A in Fig. 47, there is a conflict of currents. The flow of water from the upper water front is passing downwards to the boiler with more or less velocity, while the heated water in the boiler is being forced upwards to balance the cold water in the return leg of the secondary circulation loop. Thus one of the circulations is retarded and it is generally that from the upper water front. The water becomes overheated and steam is formed with the consequent pounding and hammering noises in the boiler and water heaters.

If the supply is from an overhead tank and an expansion pipe is carried from the top of the loop above the upper stove, the noises will not be in evidence, but the water front will not be doing satisfactory work. By using the method shown in Fig. 48 these troubles may be avoided, as there is very little resistance to the circulation from either primary or secondary sources. That from the upper heater has free and unrestricted access to the boiler, that from the lower heater equal facility, and the secondary flow leaves the boiler at the point where the hottest water is stored. Should the supply to the boiler be from the city mains proper provision must be made to guard against siphonage by using a vacuum valve.

The correct position for these has already been shown, and we need only repeat the caution to carry the cold supply to a point above the upper water front level before taking off the cold supply to the boiler branch. Thus the chance of emptying the water front unknowingly is guarded against and danger of explosion or burning of the water front practically eliminated.

Connecting Horizontal Boiler with Heaters on Different Floors.

Where it is desired to set the boiler in a basement and heat it from the kitchen range on the floor above as well as from a tank heater in the basement the connections may be made as

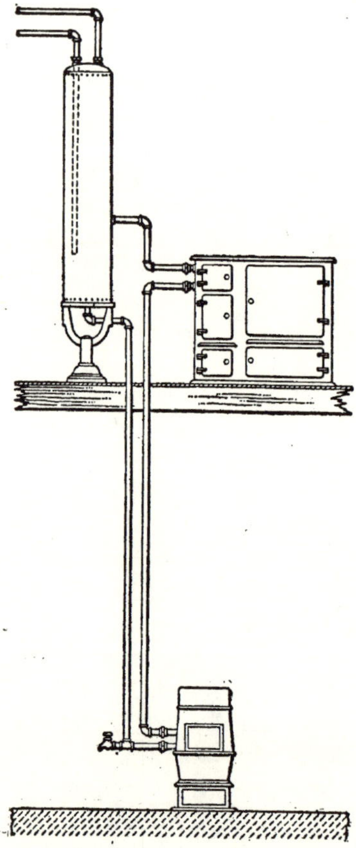

Fig. 45. Continuous Flow Connection from Two Heaters.

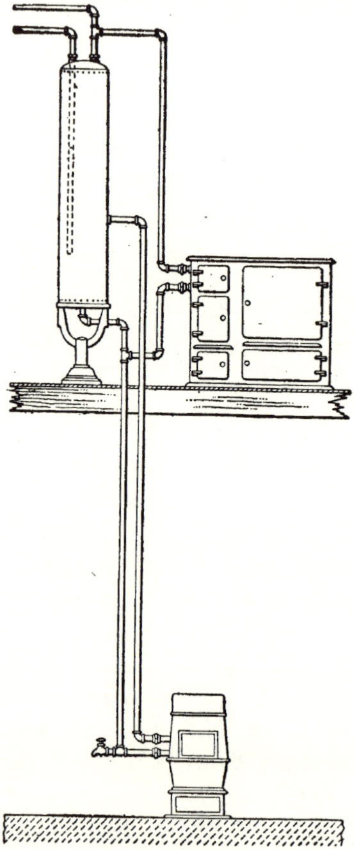

Fig. 46. Separate Connections from Each Heater.

shown in Fig. 49. These allow each heater to be used independently of the other or both at the same time without conflicting currents to retard the circulation.

The supply to the fixtures is taken from the tank by a special tapping, this being done so that the pipe may be carried to a height sufficient to prevent any chance of the water in the upper water front being drawn down below the level of the fire box should the water supply fail and there be any fixtures at a lower level than the water front. If the supply to basement fixtures were taken from the boiler without rising above the next floor level this might lead to serious trouble should the water be

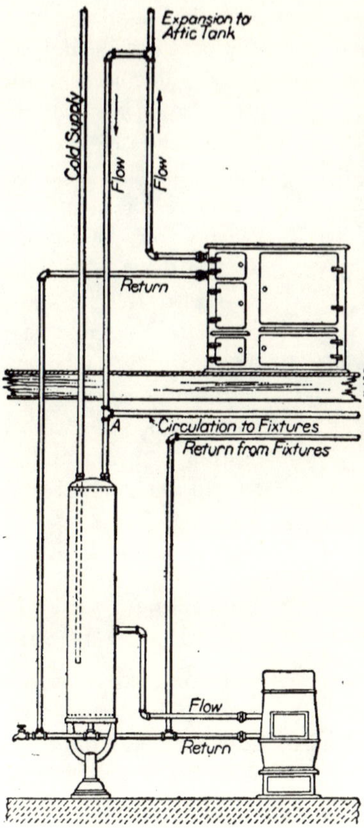

Fig. 47. Method Which Interferes with Circulation to Fixtures.

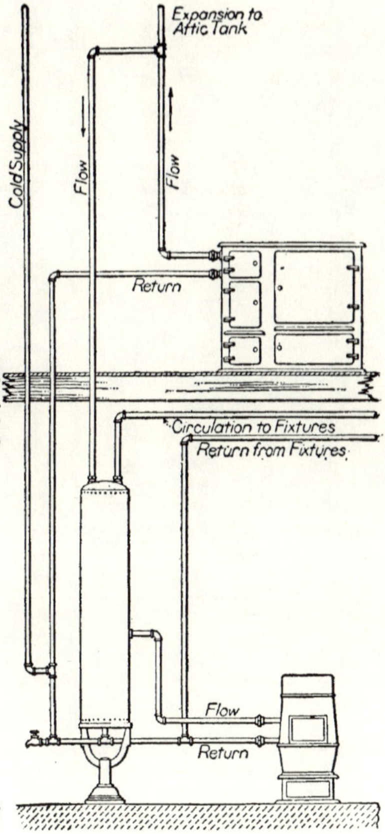

Fig. 48. Method Which Allows Free Circulation to Fixtures.

turned into the system again while the water front was red hot. If desired, the flow connection from the tank heater may enter the return connection of the water front on the floor above instead of making a separate circulation to the boiler. This simplifies the connections to some extent and gives the circula-

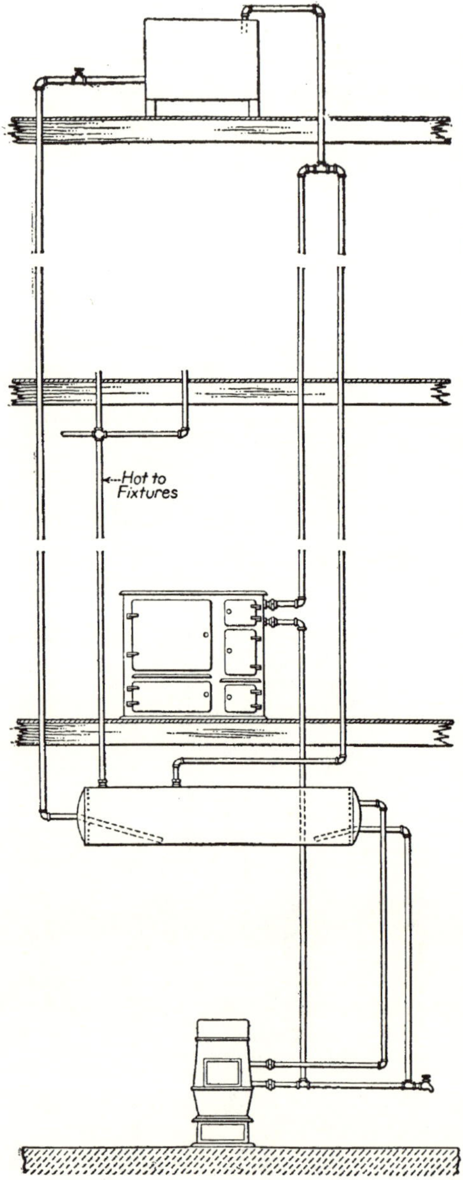

Fig. 49. Horizontal Boiler Connected with Heaters on Two Floors.

tion a continuous path but it also increases the travel of the water to the boiler with attendant loss of heat so that very little is gained. If a direct pressure supply connection to the boiler is made a connection to a fixture on one of the upper floors should be made to the highest point of the loop to remove any air collecting there. In that case also if thorough protection against siphonage of the boiler and lowering of the water line too far is desired the supply should be carried up to a high point as described in Fig. 30.

Connecting Two Boilers with Two Heaters.

There is always more or less trouble in store for the plumber who installs a series of range boilers with individual heaters when they feed into a common supply line. If any one of the heaters is for the time being out of use there is every chance of cold water being drawn from the boiler connected with it, along with the hot water from the others, and instead of hot water probably only a tepid supply is available. The equalization of pressure throughout the system immediately a faucet is opened causes this flow, and the only way to positively prevent it is to place valves on the lines to cut out the cold boiler. Needless to say, this is not to be recommended, as there is always a liability of their use being forgotten or misunderstood.

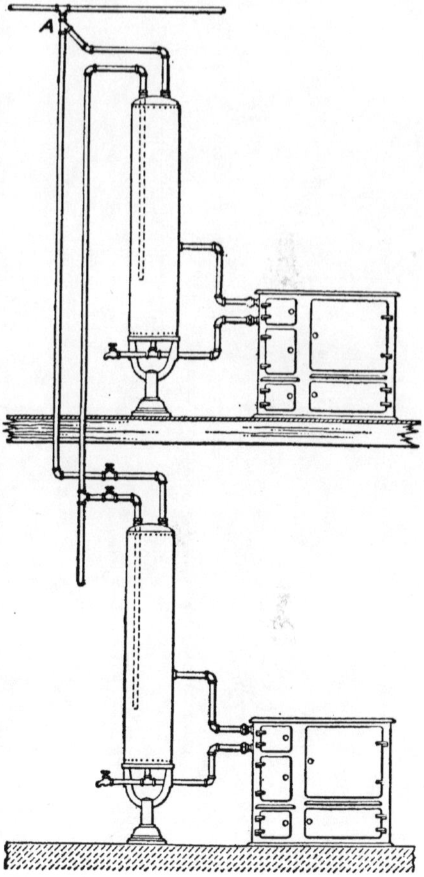

Fig. 50. Connections for Two Boilers and Two Heaters.

When the boilers are on different floors, as in Fig. 50, and the upper boiler the one generally heated, the mixing of the cold and hot water is not so noticeable, especially if a Y connection is made at A, as there is then less friction to overcome in the short connection from the vertical pipe in drawing water from this boiler than from the lower one. When the conditions are reversed and the lower boiler alone is heated, the chance of mixing is much increased and it will be next to impossible to draw water at the faucets at a temperature anywhere near that of the water in the boiler. To make a more satisfactory job the method shown in Fig. 51 is recommended. This is absolutely reliable when both boilers are in use, as the circulation between them brings the temperature of both to a practically uniform point throughout.

When the upper water front is cold more work is put upon the lower boiler, and unless the connection shown by dotted lines is also made the upper half of the upper boiler must also be heated before hot water can be drawn, and the circulation between the boilers is at the same time keeping the entire contents of the two at nearly uniform temperature. When the upper water front alone is heated this circulation is not so rapid, as the total contents of the upper boiler must be hot before any effect is shown on the lower one.

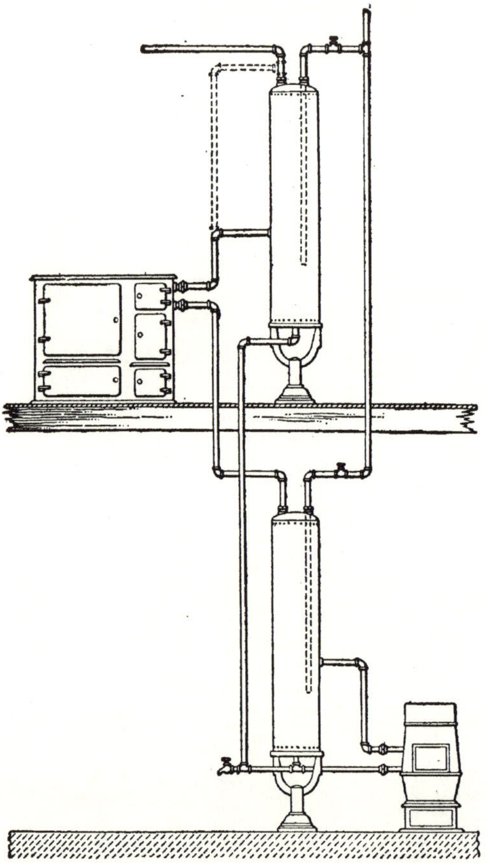

Fig. 51. Continuous Flow Through Two Boilers and Two Heaters.

The cold supply may be omitted from the upper boiler or the valve kept closed if desired to insure the whole contents being available at the faucets. This style of range connection is more generally used than that shown in the next illustration, Fig. 52, and certainly has much to recommend it on the score of simplicity as well as efficiency. About the only weak point in the system is the passing of the flow from the lower boiler through the upper water front, as it is thus cooled off to some extent. When the connection is made as in Fig. 52 this is overcome, as the flow from the lower boiler enters the upper boiler through the side inlet in combination with the flow from the water front in the stove. It can be continued to the top tapping as in a quick heating connection if preferred, or the flow from the upper water front can enter the boiler in this way, thus making the two circulations distinct. When they both enter through the side opening the piping must be enlarged at the junction to the largest size possible with the tapping provided. This lessens the chances of retarding the flow from either of the heaters by that from the other, a condition which makes for overheating, rattling and pounding.

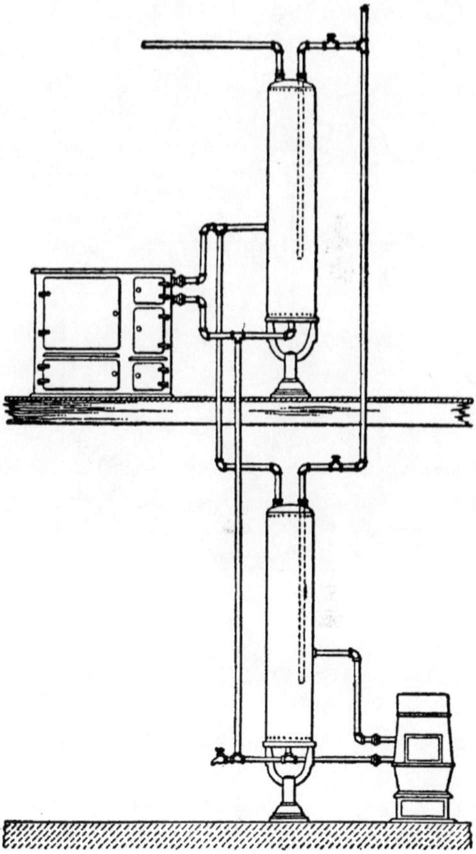

Fig. 52. Connection to Allow Use of One Stove Only if Desired.

As in the previous example, the use of a cold supply pipe in the upper boiler is optional, as when the supply comes from the lower one alone every drop of hot water can be displaced before cold is drawn. Objection to this method of supply is sometimes made on the score that the cold water shoots clear to the top of the boiler and cools it off, but this is only the case under heavy pressures and ought only to be considered when such obtain. Even then the effect of the flow may be over-estimated.

Joint Hot Water Supply for Two Flats.

The two methods of connecting boilers on different floors with ranges on the same floor as the boilers, but which admit of heating the entire system with either as may be convenient, are shown in Figs. 53 and 54.

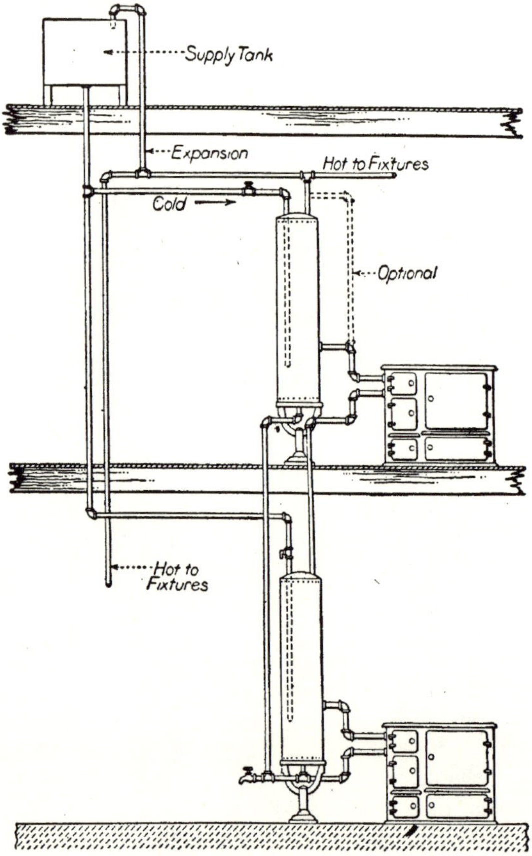

Fig. 53. Joint Hot Water Supply for Two Flats.

In Fig. 53 is shown a 30-gal. boiler, connected up so that all the hot water from the lower boiler will pass through it before it is drawn at the fixtures. This has certain advantages and also disadvantages. If the upper range is being run alone, hot water will be readily drawn at any fixture without cooling by mixing with the cold water from the lower boiler. If both boilers are being run, all the hot water in both can be drawn if desired, and the circulation is free and continuous through the whole system. No valves are necessary on the hot supply pipes. The valve on the cold supply to upper boiler may be kept open

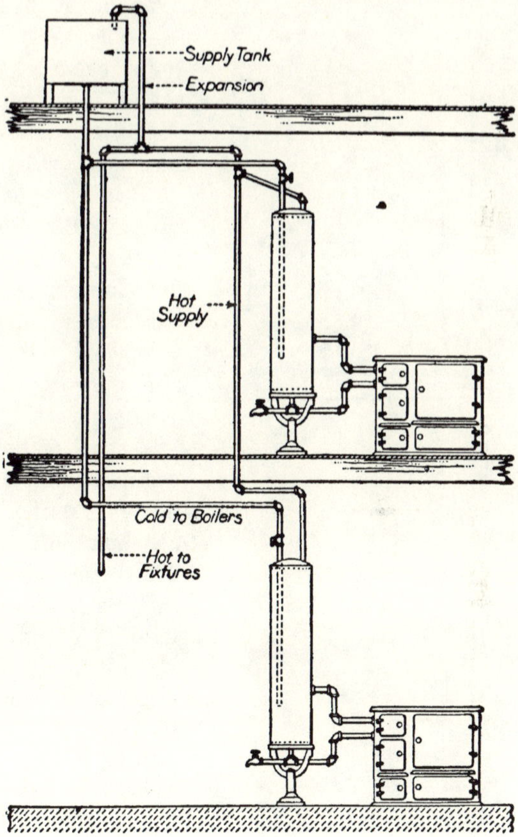

Fig. 54. Two Boilers and Heaters Connected to a Common Supply Line.

or closed, as desired, as the pressure from the tank is low and a supply entering the upper boiler through the return would not have velocity enough to cool the water in the upper part of boiler.

When the upper range is out of use, one disadvantage is apparent: The lower range has to heat the water in the lower boiler and also that in the upper portion of the other before hot water can be drawn at the fixtures. To get over this, fit a quick-heating connection to the upper boiler, as shown by dotted lines. This stores the water in the upper portion first, and hot water is thus more quickly available at the fixtures.

The tank must be high enough to take care of the expansion, or a continuous stream of hot water will pass into it from the expansion pipe; and, also, there will be a liability of drawing air through this pipe when a faucet is opened, with attendant gurgling and unsteady flow.

Fig. 54 shows a common method of connecting two boilers to a common supply pipe. The principal objection to this system is the liability of mixing cold water from a boiler which may not be in use with the hot from the other when drawing from any of the faucets. Valves may be fitted on the supply pipes, but the practice is objectionable and the system shown in Fig. 53, although a little more elaborate, is much to be preferred.

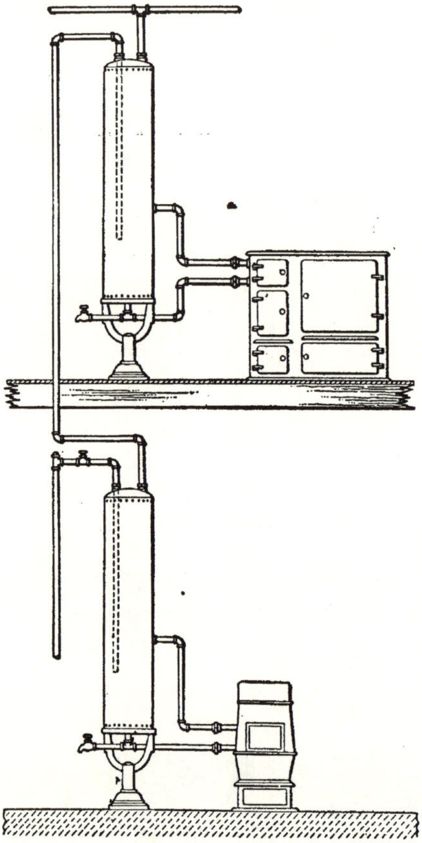

Fig. 55. An Unusual Method of Connecting Two Boilers and Heaters.

Unusual Double Boiler Connection.

Local conditions often call for the use of special methods of making connections from two boilers and heaters to the hot-water supply lines. The two styles previously described can be used successfully in most cases, but occasionally conditions are such that a simpler method may be adopted and a considerable saving in time and material effected without in any way impair-

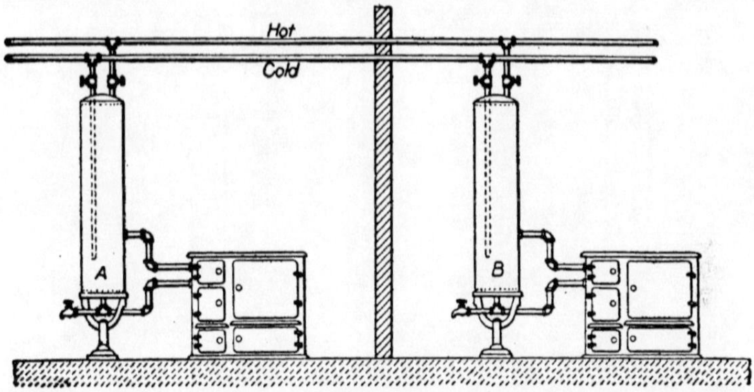

Fig. 56. Common Method of Connecting Two Boilers on Same Level to Common Supply Line.

ing the efficiency. Such a method is shown in Fig. 55. The only example of this connection that the writer recollects was put in by a hotel man to avoid changing the existing piping to suit one of the more regular methods of installing the auxiliary boiler and heater in the basement. In his case the upper fire is continuously used and therefore the function performed by the lower heater is simply preheating, and that of the lower boiler storage of the pre-heated water until the opening of a faucet allows it to pass into the upper boiler.

In some cases overheating of this upper boiler would certainly take place and each time the opening of a faucet relieved the pressure considerable noise would be made by the steam, but where so much water is required and the drawings are so frequent, as in this instance, this trouble does not arise. No doubt this job would be a poor one where the upper fire is not in continuous use, as then absolutely no hot water could be drawn. The connection should, therefore, never be made without careful consideration and assurance that the right conditions prevail.

Suggestions on Double Boiler Connections.

In Fig. 56 is shown another example of the common style of connecting two boilers to a common line of supply piping, but both boilers are on the same floor in this case. Here also one may expect trouble; in fact, the drawing shows an actual installation that was found unsatisfactory and remodeled. The

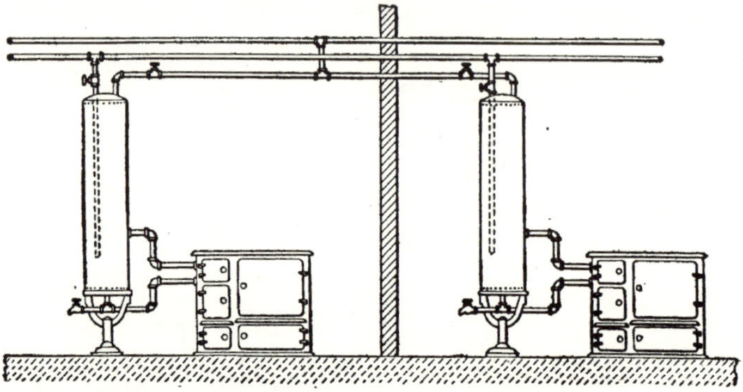

Fig. 57. Method of Balancing Flow from Two Boilers.

trouble here lies in the liability to draw almost exclusively from the nearer boiler, whether it be hot or cold. The only way in which the hot water in the other boiler can be drawn is to close the control valves on the cold one and this is an undesirable and troublesome condition to impose on the user. A little better service is gained by taking the outlets from the boilers to a point

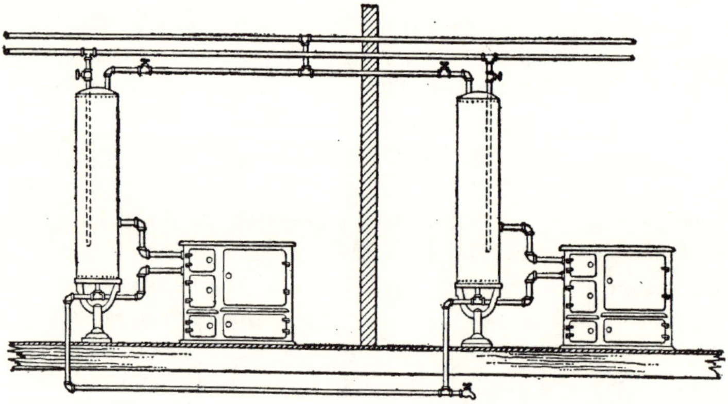

Fig. 58. Connections for Circulation Between Two Boilers on Same Level.

midway between them, as shown in Fig. 57, and then connecting into the main hot water line. This equalizes the friction so that an equal amount of water is taken from each boiler. So long as ach is near the same temperature the flow at the satisfactory and near the same point as obtains but should one of them happen to be much cooler a considerable reduction must be expected. These must have the attention of users, and if a steady water is demanded attention must be given to rs equally.

ows the same connection with a circulating pipe s a comparatively equal temperature of water in whether one or both heaters are in operation.

CHAPTER VII.

Supply Connections and Distribution.

The most common system of distribution to the different fixtures in small residences, flats and tenements is to carry one main supply pipe from the boiler and from that take branches at convenient locations for the different fixtures. This system has objectionable features when the work is such that long runs of pipe are required or such that branch connections are difficult to make in a location that will allow of a stop cock for each branch being fitted. In the first case a considerable quantity of cold water has to be drawn each time a faucet is opened before hot water is available. This is wasteful both of water and heat as all the water that is drawn from the boiler and which stands in the pipe from the faucet to the boiler is cooled each time the faucet is closed again and must perforce be wasted by the next user. In the second case it may be inconvenient to turn the water off the whole system because of some fault on one line, yet no provision can be made to avoid this owing to the manner in which the pipes have been carried.

Again, if the pipes are turned down into the basement as shown in Fig. 59, and which is a very common method of pipe fitting, there is opportunity of air collecting in the loop and retarding if not altogether stopping the supply to the fixtures. This is more likely to occur when a low head of water is carried in the system, as a high pressure will force any air collected in the loop to the faucets each time they are opened. To avoid this annoyance the pipe may be run as shown in Fig. 60, dropping the branches to cellar fixtures and to those on the same floor as the boiler while the others are taken from the top side of the main pipe, thus providing opportunity for air to escape freely. Stop cocks should be fitted on each branch and if these are of the stop and waste pattern, allowing the water left in the pipe to drain back when the stopcock is closed drip pipes may be fitted as shown in Fig. 61, thus avoiding any leakage of water on the floors or walls while the pipe is being emptied.

Supply from Water Tables.

Another plan often followed is one which requires considerably more pipe and more labor but the results justify the extra expense as a rule. This is the system where the hot and cold supply mains are taken to a water table and the branches taken from a distributing header. This makes an extremely neat job which also has the recommendation of bringing all the control valves to one point where they can be conveniently la-

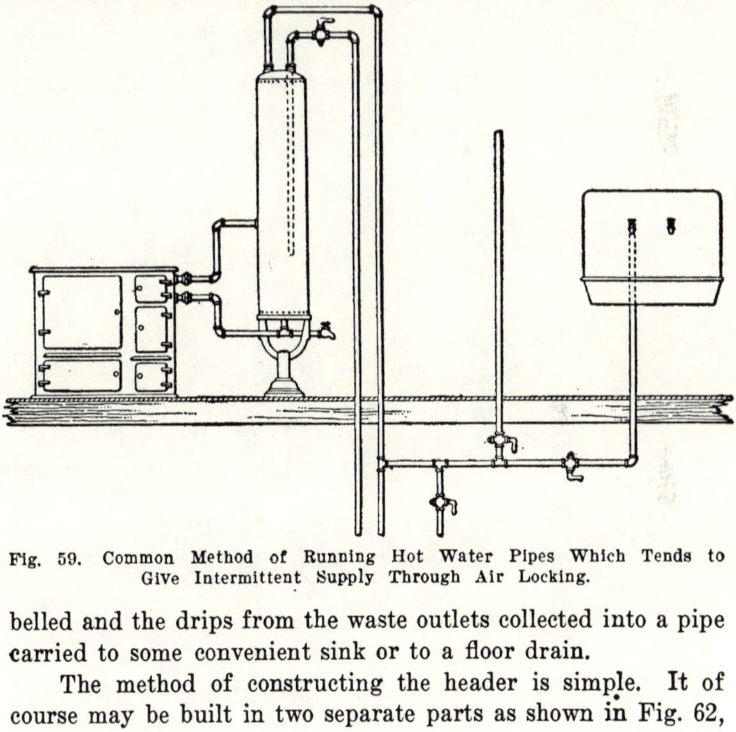

Fig. 59. Common Method of Running Hot Water Pipes Which Tends to Give Intermittent Supply Through Air Locking.

belled and the drips from the waste outlets collected into a pipe carried to some convenient sink or to a floor drain.

The method of constructing the header is simple. It of course may be built in two separate parts as shown in Fig. 62, but better appearance is secured and also more compactness if it is built as shown in Fig. 63. All the difference is that instead of plugging the end tee the two end tees of each header are connected by a solid nipple. This nipple is easily made if the supply house cannot furnish them. Take a brass nipple of the same length as those fitted between the rest of the tees and file down inside it as far as can be reached until a perfectly clean surface is secured. Then tin it, using cut acid and heating the nipple over the fire pot so that the solder will run well. If the nipple is heated well it will only be necessary to rub the solder on it, with an application several times of the acid brush. When the inside is thoroughly tinned pack some paper in it, leaving about ¾ in. at each end. Then fill this space with solder, heating the nipple with a torch or over the fire pot to make the solder flow freely. The stopcocks should be fitted at the same height on the vertical branch to preserve a symmetrical appearance and each nipple should be of the same length. The main pipes

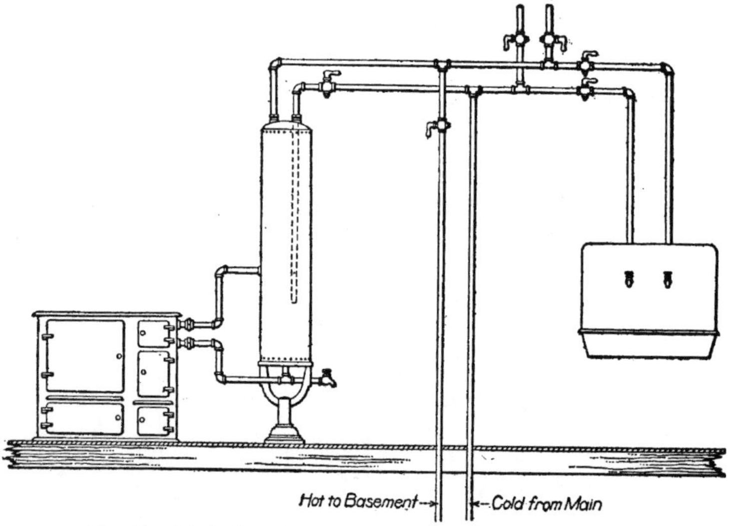

Fig. 60. Method of Running Hot Water Pipes to Fixtures to Avoid Air Locking.

entering at the end should have a plug or petcock fitted in a tee to admit of draining the line. All horizontal pipes run in a cellar should be run on as equal a grade as possible to secure good appearance and should either drain toward the header or to some plug in the end of a tee. Adjustable hangers make the fitting of these pipes much easier, as the pitch can be adjusted to suit any condition.

Circulation of Hot Water to Fixtures.

When the plumbing fixtures to which hot water is supplied are at any considerable distance from the range boiler, the necessity of drawing a large quantity of cold water before any hot water is available becomes somewhat objectionable. This consequence is avoidable and the waste of water is unnecessary, as in nearly every case the hot water may be made available almost immediately at each faucet by a proper system of circulation between the boiler and the fixtures. This is generally described as a "secondary" circulation, distinguishing it from that established between the boiler and heater, and is arranged in a manner that will not affect the primary circulation.

There should be little in any ordinary building to prevent

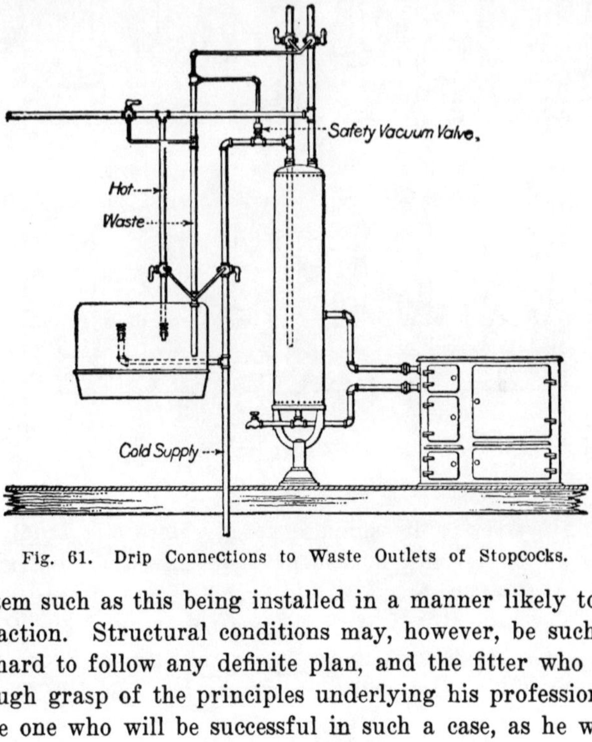

Fig. 61. Drip Connections to Waste Outlets of Stopcocks.

a system such as this being installed in a manner likely to give satisfaction. Structural conditions may, however, be such that it is hard to follow any definite plan, and the fitter who has a thorough grasp of the principles underlying his profession will be the one who will be successful in such a case, as he will be able to adapt his design to the building by taking advantage of more than one method of securing circulation. For instance, it is often very difficult to secure sufficient pitch between the floor and ceiling to make a circulating loop work well, or objection may be made to placing long runs of piping in such positions. In such a case a combination of loops with a falling circulation may be satisfactorily installed. There are many other combinations, some of which will be illustrated, and all of which ought to be within the knowledge of the plumber who undertakes hot-water supply on any other than the simplest scale.

An important consideration in this work is the provision of sufficient pitch to the pipes at all points and avoidance of air pockets. Provision must always be made, by having a fixture connection or other means of relief at the highest point in the

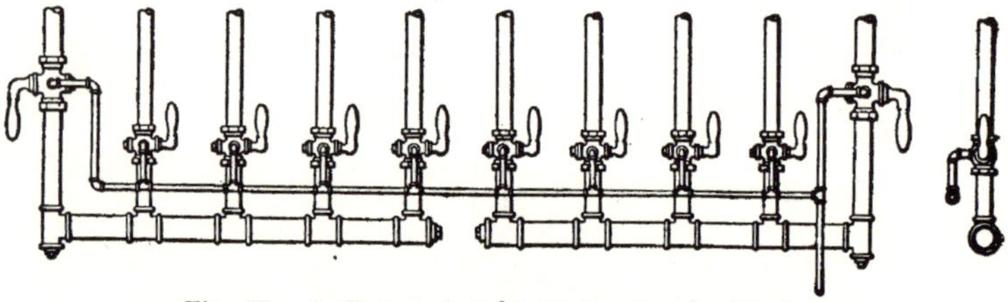

Fig. 62. A Hot and Cold Water Supply Header.

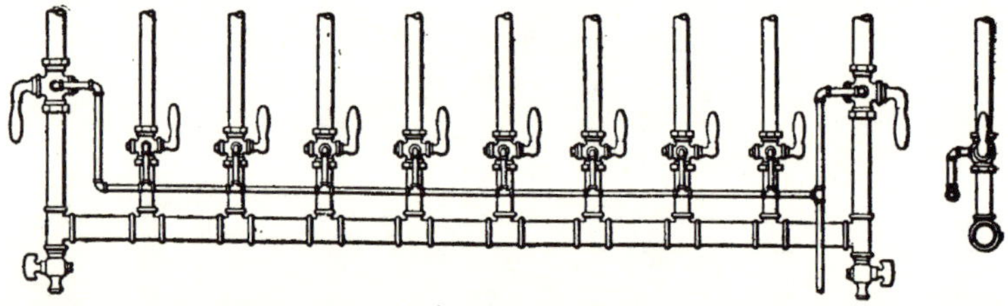

Fig. 63. A Header with Solid Connection Between Hot and Cold Ends.

circulating system, to allow air to escape. Fig. 64 shows a circulating system in its simplest form. Here the supply pipe simply forms a continuous loop which travels around the fixtures, branches being taken off at the points nearest to the fixture. The pipe pitches upward from the boiler to the last tee before it drops back to act as a return pipe. Thus any air collecting in the pipe will be relieved each time this faucet is opened. Should a very long run be necessary to reach the last tee, the high point may be made at any of the other tees, the point to be remembered being that the pitch to this point must be continuously up and then continuously down, so as to avoid pockets that would impede circulation.

In Fig. 65 we have a somewhat different arrangement of piping supplying the same fixtures; each bath room having an independent branch circulation. At the point marked A in this illustration the better method of making the return connection is shown. This is better than that shown at B, where the space available is limited, as there is then no risk of having the circulation stopped by sagging pipes forming air pockets. The loop system may not offer any particular advantage over the one first described, but it may be that conditions are such that it is the only one that can be satisfactorily adopted, and in a job of considerable size is generally to be preferred for the reason that the circulating path will probably be shorter and the supply consequently hotter.

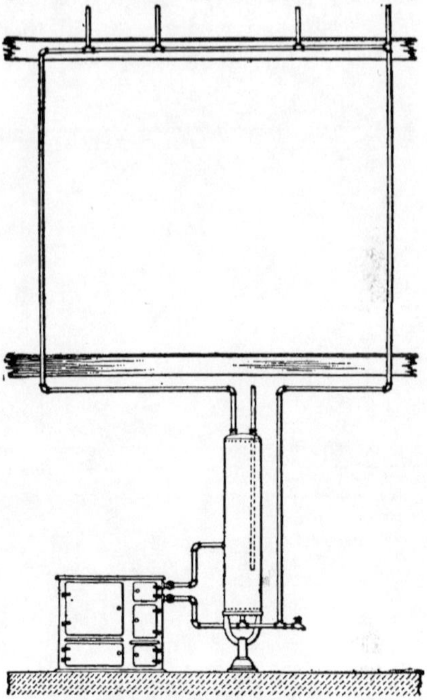

Fig. 64. A Simple Circulating System for a Cottage.

In the system shown in Fig. 65 there is sometimes a danger of drawing cold water through the return at the last fixture on the loop, as the friction on the long length of pipe that the water has to traverse is considerable, and the last fixture being nearer to the boiler on the return line than on the flow, the cold water is somewhat liable to back up when the faucet is opened. To remedy this a check valve is often fitted, but this acts somewhat as an impediment to free circulation, and it is a better plan to take the branch to this fixture from the flow pipe after it leaves the boiler, making a separate loop for this alone, according to the method shown in Fig. 66.

Swing check valves are the better pattern to use when it is necessary to have them, but no matter how light they are they are liable to become set either closed or open and so defeat the purpose for which they were fitted. By setting a swing check in an inclined position there is no interference with circulation however sluggish, yet the check will close under a reverse flow.

A Novel Hot and Cold Water Supply in Combination.

The arrangement of piping shown in Fig. 67 is rather more of a novelty than an example of standard practice, but it embodies some ideas which are easily applied and the connections are simple and easily adapted to various conditions.

The equipment as shown in Fig. 67 consists of a 20 gal. extra heavy galvanized boiler, hung from the cellar ceiling, connected to a 6-in. water heater in a steam-heating boiler. The hot-water pipe, ¾ in., runs to the basin sink, with a stop under the basin and a ⅜-in. circulating pipe is taken out below the stop and runs back to the lower connection between the boiler and heater. The water pipe runs to the ice chest in ½-in. lead pipe, which is coiled closely, covering the entire bottom and then rising and running to the basin and sink, with a stop near the point of connection to the hot-water pipe. Over the pipe coil in the bottom of the box there is a bottom of wooden slats, on which the ice rests. The waste pipe from the ice box extends through the bottom of the box with a little extension piece, so that the pipe coil rests in and is surrounded by the melted ice water. The drip empties into a closet tank on the lower floor. An entire cake of ice is placed in this box and covered with blankets. This usually lasts about a week or eight days, excepting in extremely hot weather. The water at the faucet is al-

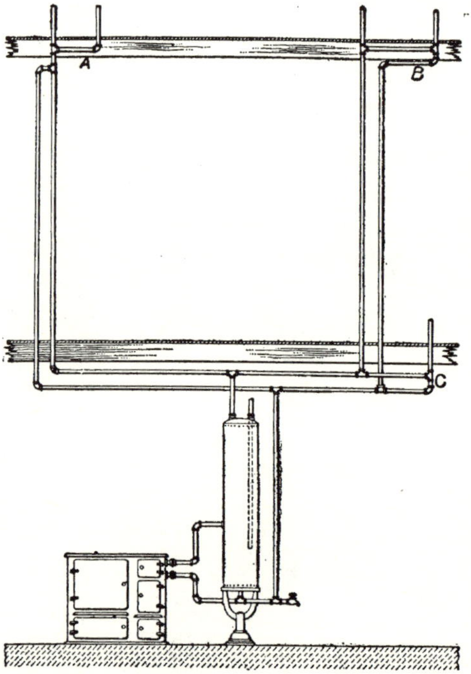

Fig. 65. A Circulating System with Branches Planned to Avoid Air Locking or Sagging Under Floors.

ways cold and refreshing. In the summer the hot-water stop is closed and the ice-water stop opened, while during the season when the steam boiler is running the ice-water stop is closed and the hot-water stop opened, thus providing a supply of ice water in summer and hot water in winter with one system of piping. The ice chest is 3 x 2 x 2 ft.

Features of Circulation in a Cottage.

The example shown in Fig. 68 contains the problem of heating a boiler with a door intervening between it and the range and that of making a circulation to the heater in the cellar while providing a supply of hot water at the fixtures immediately on the opening of a faucet.

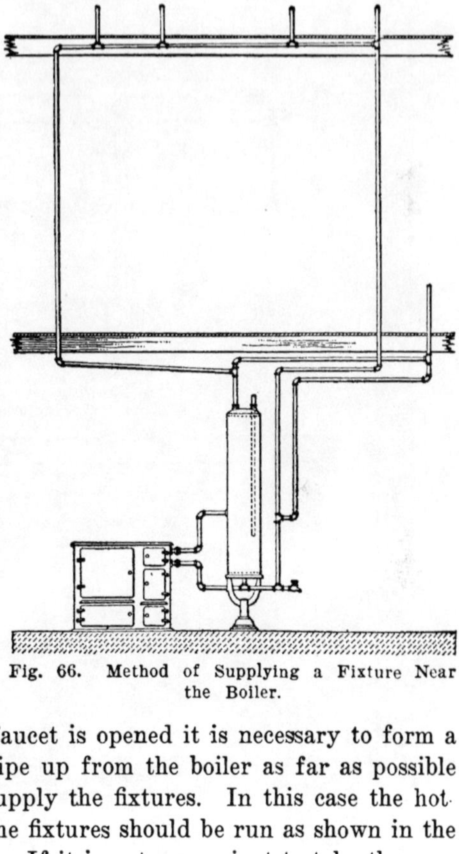

Fig. 66. Method of Supplying a Fixture Near the Boiler.

To supply hot water to fixtures below the level of the boiler or on the same floor as the boiler so that the hot water can be drawn as soon as the faucet is opened it is necessary to form a loop by running the pipe up from the boiler as far as possible before descending to supply the fixtures. In this case the hot water supply pipe to the fixtures should be run as shown in the accompanying drawing. If it is not convenient to take the supply pipe across the ceiling it can be run as shown by dotted lines. If this method is used the pipe should be carried even higher before descending. The necessity of having an air valve at the top of the loop is dependent upon the amount of pressure carried in the pressure tank. If the system is run under low pressure the air valve should be used so that it can be opened to let air escape when the pipe becomes air bound sufficiently to stop the flow or circulation.

In Fig. 69 is shown a system for utilizing separate range boilers in an apartment building. In addition to five floors of

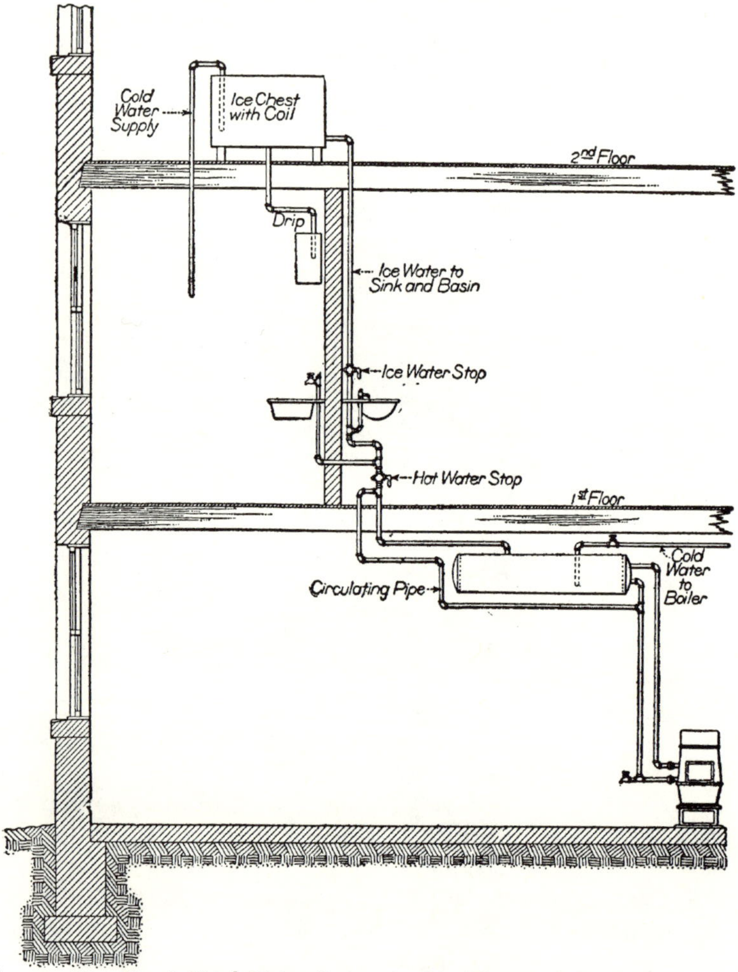

Fig. 67. A Novel Piping System for Hot Water and Ice Water.

the building above ground, the range boiler in the janitor's apartments in the basement where the heater is located, is also supplied with hot water. A small round boiler having a 15-inch grate, and rated to carry 200 square feet of direct radiation and having a capacity for heating 100 gallons of water per hour from 40 to 212 degrees, is connected with the six boilers, as shown, each one of which has a capacity of 30 gallons. A 2-inch flow main is carried up from the little heater and 1-inch branches are taken to the side connections on all the boilers on the upper floors. A tee is used at the side connection to receive

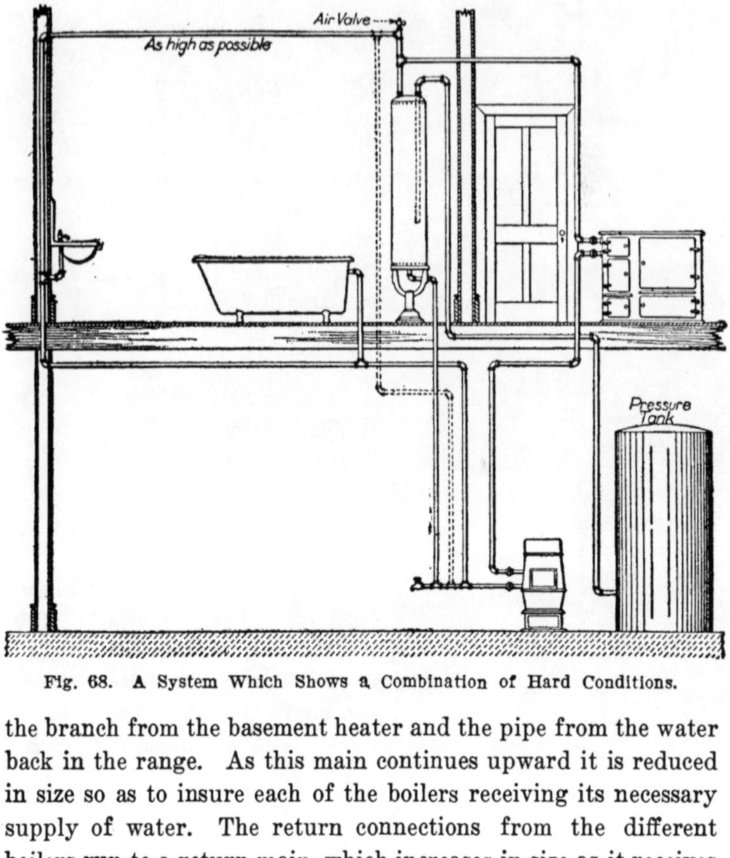

Fig. 68. A System Which Shows a Combination of Hard Conditions.

the branch from the basement heater and the pipe from the water back in the range. As this main continues upward it is reduced in size so as to insure each of the boilers receiving its necessary supply of water. The return connections from the different boilers run to a return main, which increases in size as it receives the different connections and is carried to the boiler. The boiler in the basement is heated by means of a 1-inch branch, which is connected to the top of the boiler at the hot water service outlet, and the return from this boiler is carried to a separate opening in the base of the water heater.

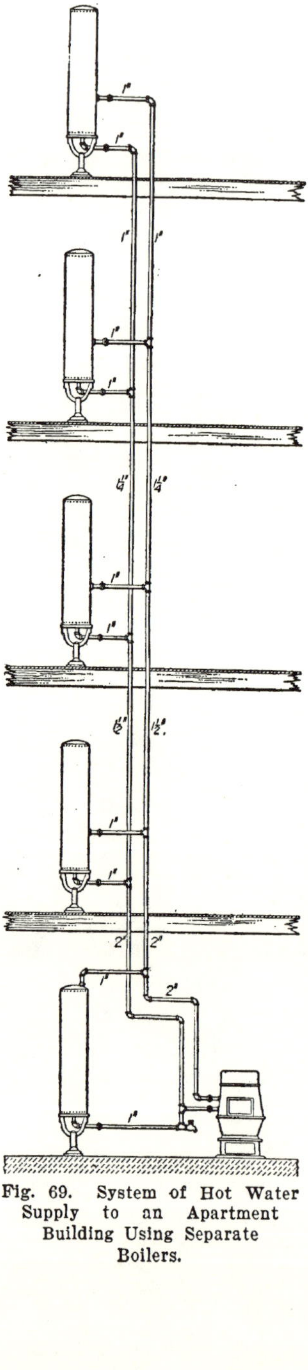

Fig. 69. System of Hot Water Supply to an Apartment Building Using Separate Boilers.

This method of connecting allows each tenant to use a coal fire when an extra supply of hot water is required for laundry or other purposes. Each boiler is treated as a radiator and piped accordingly. It will be noted that the six boilers have a capacity for holding 180 gallons of hot water, while the heater is only rated to furnish about one-half of this quantity. This seeming lack of power is offset by the storage capacity of the different boilers, and the allowance is made for the fact that all of the hot water in a boiler is seldom drawn from it at one time.

Piping Systems in Large Residences.

The system of hot-water supply circulation that is most suitable for tall buildings of the office type, or apartment buildings supplied from a common heater, is that known as the overhead or falling circulation. This system is also eminently suitable for buildings of less pretensions, especially if there is an attic in which the various branch lines radiating from the main risers can be carried to a point directly over the fixtures they are to supply. When this is so it is possible to install the piping in a manner which calls for a minimum risk of damage from leakages, as it is necessary to put only a very little of it under the floors of bath rooms or toilet rooms. Where these are situ-

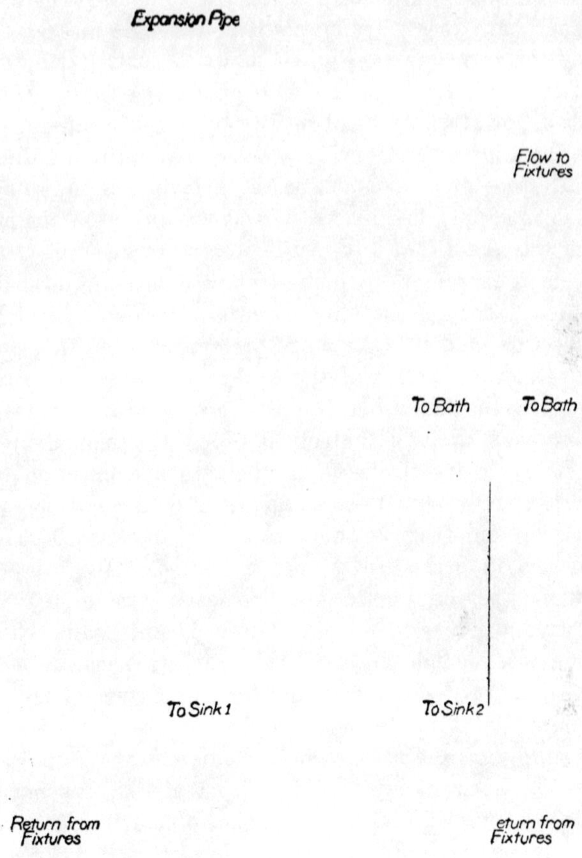

Fig. 70. Drop Feed Circulating System for a Large Residence.

ated over rooms where a leakage might cause havoc with furnishings this point is of importance, and also where the floors of such bath rooms are of tile it is desirable to keep the piping away for easy access. Fig. 70 shows an installation of this type in a large residence, the boiler supply being from a copper lined tank in the attic. As will readily be seen the difference between this installation and one in a large apartment building is principally that of proportion, as the principle is the same in each.

The supply to sink No. 1 is shown taken off the main riser. This was done for convenience, the more general custom being to carry this pipe to its highest point before any branch is taken from it. There is nothing essentially wrong in making such a connection, and it is certainly preferable to making one so near the return connection of the boiler that there is a possibility of reversing the flow when the faucet is opened and drawing cold water back through the return pipe. Occasionally it is found necessary to insert checks on the return connection to prevent this happening, but this should never be done if the system will work without them as they are always liable to become clogged and remain permanently open or closed. If a fair velocity of flow can be induced in the circulation system the chances of drawing the reverse way are lessened, and again, where such a liability is thought to exist, the connection can be made to the drop line at a point above the level of fixture. The connection for sink No. 2, for instance, might have been made through the tee supplying the bath room immediately above it.

The horizontal piping in a system like this must be very carefully pitched and the expansion pipe connected to the highest point so that air will not collect. It may be interesting to the student of hot-water supply work to compare this installation with one in a similar house in England. It will be seen that in the English installation, Fig. 71, the range is of the built in type, the circulation pipes being carried down behind rear plates to the water back. The pipes pass through the wall immediately above the mantel shelf to the boiler which is of a shorter and wider build than those generally used in the United States. It is placed on brackets of wrought iron and in a corner of the kitchen above head level. As it is not large enough to serve the house alone an auxiliary is placed in the attic and a circulation is maintained between the two. A falling circulation is provided

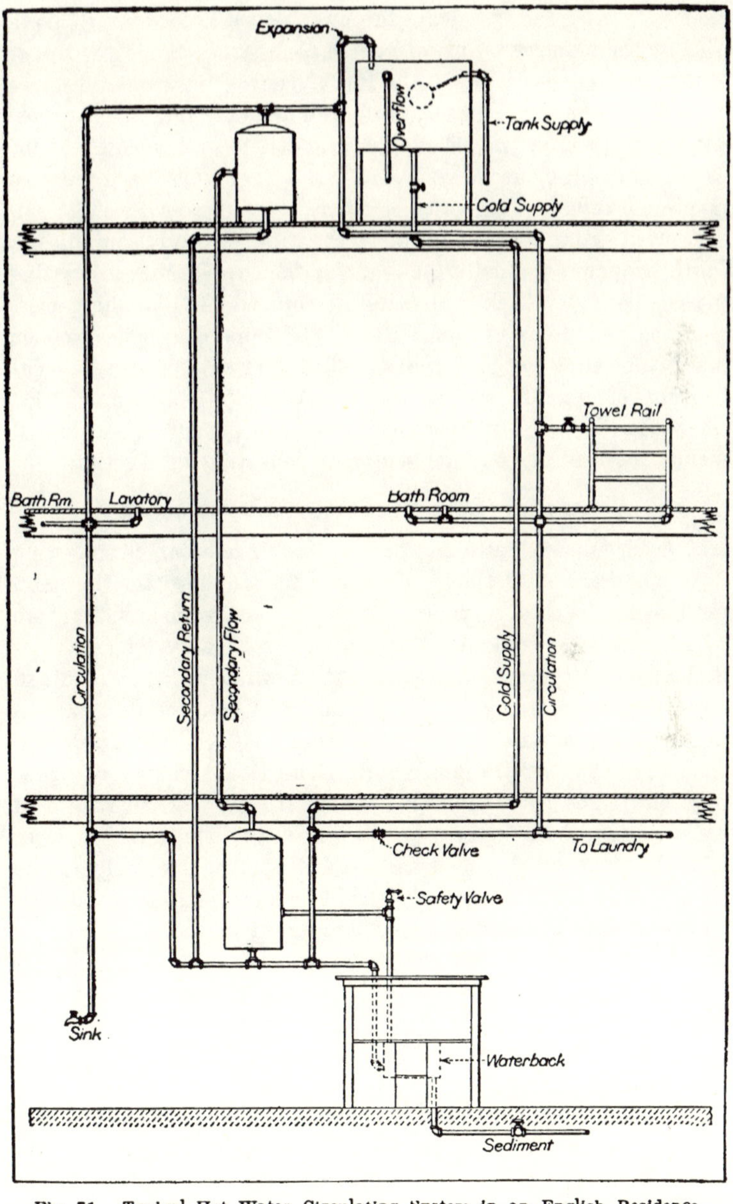

Fig. 71. Typical Hot Water Circulating System in an English Residence.

to each bath room, short branches from the drop lines being all that it is necessary to draw through before hot water is available at the fixtures. This of course commences above the upper boiler, and is connected to the top of that.

A feature of the system is the towel rail, which is built from 1¼ in. brass pipe, nickel plated, and which is connected to the circulating pipe through a valve. A safety valve is fitted just above the range. This valve is generally of the dead weight pattern with ground seats and in most cities its use is compulsory. Hard copper is the material used for the supply and circulating pipes, the water back also being of copper. For all high class work this metal is recommended. The tank is a large wooden one lined with sheet lead, walls of 6 lb. and bottom of 8 lb. The overflow comes through the bottom. A ground spigot and socket connection is wiped in the bottom so as to provide a convenient means for flushing out the sediment collecting in the tank.

Circulating Loop on Same Floor as Boiler.

A circulating system which is constructed in a somewhat different manner is shown in Fig. 72. In this case the main circulating loop is carried along on the same floor as that on which the boilers stand. This is often necessary by reason of difficulties in the building construction which will not admit of any other style being used. By taking a connection from the highest part of the loop any air that may collect is removed each time that fixture is used and there is less chance of drawing cold water back through the return pipe with this form of construction than if the return pipe were to be carried down into the floor below the boilers and connected by returning to them below the water line. It will be noticed that the boilers in this system are heated by a water front in the kitchen range and by a laundry heater and that valves are placed on the connections to the latter. This is done so that the supply may come entirely from the kitchen range except on such days as the laundry is in operation when the additional call for water is such that the extra power is necessary. If the valves were not closed there would be little chance of drawing hot water at the laundry fixtures or even at the bathroom nearest the boiler in the laundry, as the flow will always proceed from the boiler offering the least resistance and that one is nearer than the one in the kitchen. Therefore the valves are closed and any mixing of hot and cold water from the two boilers is eliminated.

To safeguard the boiler in the event of a fire being lighted when the valves are closed a safety valve is fitted, a tee being inserted in the outlet connection between the valve and the boiler. The use of valves can occasionally be dispensed with if it is possible to take the connections for the fixtures from a point in the circulating loop that will be equi-distant from each boiler. Thus

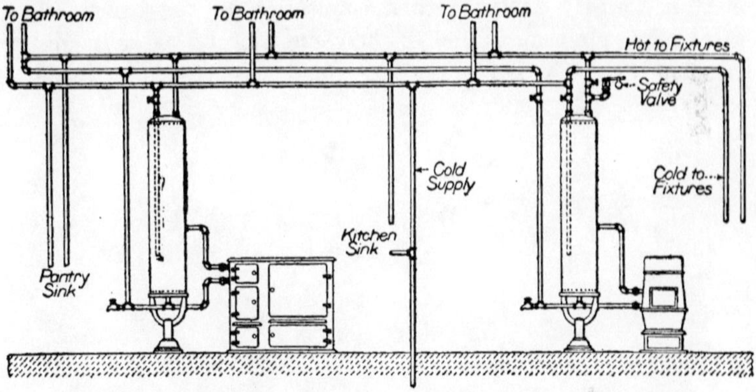

Fig. 72. A Loop Circulating System with Two Boilers and Heaters.

the friction is equalized and the flow will come as easily from one as the other. When this is done it will be found that the flow will come from the heated boiler and that there will be almost none from the other. This is probably due to the difference in density of the hot and cold water and the balance being in favor of the heated boiler, the flow will proceed from there.

Circulating Water to Fixtures on Level Below Boiler.

When it is desired to supply fixtures on a lower level than the boiler from a circulating system it will sometimes be necessary to use a light swinging check valve on the return so as to prevent the drawing of cold water back through the return pipe. If the supply to the fixtures is large and a valve is placed in the return pipe so that it can be set to pass only enough water to maintain a circulation the chances of drawing cold water at the fixtures by reversing the circulation will be somewhat reduced. It is also possible by designing the piping properly to reduce the friction on the pipe between the branch and the fixture so that it will be easier for the water to flow in the desired direction than from the return connection to the boiler, and this is a more desirable arrangement than the use of valves or checks. Wherever possible the elimination of elbows and sharp turns between the branch and the fixture will be an aid to ensuring satisfactory flow. It must also be remembered that the supply must come from a great enough elevation to ensure motive force enough in the descending column to overcome the difference in density at the return connection to the boiler. The same principle that applies to the connection of water fronts on the floor above boilers must be applied here.

CHAPTER VIII.

Hot Water Circulation in Large Buildings.

The hot water supply to apartment houses and other buildings requiring hot water in large quantities is somewhat of an engineering proposition when the building may be twelve or more stories in height. Up to this height, however, the departure from that followed in buildings of less proportions is not great. When the house tank is placed on the roof and there are fixtures on each floor and in the basement the flow at the different levels is unequal owing to the great difference in pressure at the different floors. This is corrected in some measure by the provision of valves on the supply pipes to each fixture and the excessive flow at the lower fixtures is thus checked and the annoyance caused by spattering avoided. Where there is a pent house and a sub-basement and the boiler and pumps are placed at the lowest level the pressure carried there is somewhat high, and in a building of twelve stories, not including the basements and pent house, the pressure will approximate 90 lbs. at the boiler. This pressure is not greater than is carried in many city mains and is not a disadvantage otherwise than that it causes a somewhat heavier pressure on the steam coil in the boiler than the cast brass elbows used in the construction of coils will stand unless these are made extra heavy. In some instances the act of forcing the threaded brass pipes up to the end of the thread with the idea that the joint is being made more secure has caused the fittings to spread and the pressure of water being heavier than that of the steam, water has entered the steam pipes and led to trouble. To avoid this many engineers are using wrought iron galvanized fittings with brass pipe, preferring to take the chance of these corroding but making provision in their construction for the easy replacement of coil headers, elbows, or return bends. The system of circulation most commonly used in such an installation as a twelve story apartment or office building is generally of the drop feed type. In this system the flow pipe is carried clear to the top of the building and a vent pipe, shown in Fig. 73, carried to a point over the house tank, is taken from the highest point. From there the main return pipes are taken and the different lines which are to supply the fixtures are taken off at such points on these return pipes as may be convenient. The drop lines are carried down through the building sending off such branches as may be required and on reaching the basement are connected into a main return line. At the foot of the vertical line a control valve is usually fitted and in addition a check valve is placed behind this so that when faucets are opened there will be no chance of reversing the circulation by drawing water back through the return pipe. As there is considerable expansion on the risers of a building of this height it is usual to provide against damage resulting from it by putting in the line a swing joint constructed of ells and short pieces of pipe as shown in the illustration. When the pipe lengthens with an insertion such as this all that happens is that the short piece of pipe turns slightly in its fitting, allowing the lateral pieces to rise and thus take up the movement without strain on the pipe or fittings. The branch pipes to the fixtures may or may not have control valves on them but it is good practice, and one usually followed to have valves on the exposed supply pipes to each fixture. If the branch pipes have to run any distance laterally it is well to make a return connection to the supply so that a large quantity of cold water will not have to be drawn off each time a faucet is opened. The method of doing this is indicated in the illustration also. A valve may be placed on each leg of the branch or a check valve only on the return connection, the object being to prevent water flowing through it should it be necessary to close the branch valve for any purpose.

Distribution From Rising Mains.

When a rising supply system is installed the piping is practically the same as that illustrated for the drop feed system. In this case, however, the proportions are reversed and the sizes of the pipes would be reduced as the higher elevations were reached and the number of faucets to be supplied reduced. The connections at the tank are made in the opposite manner also, as may be seen by reference to Fig. 74. The branches are taken off the main flow branch and after passing all of the connections to the various apartments are collected into a common return

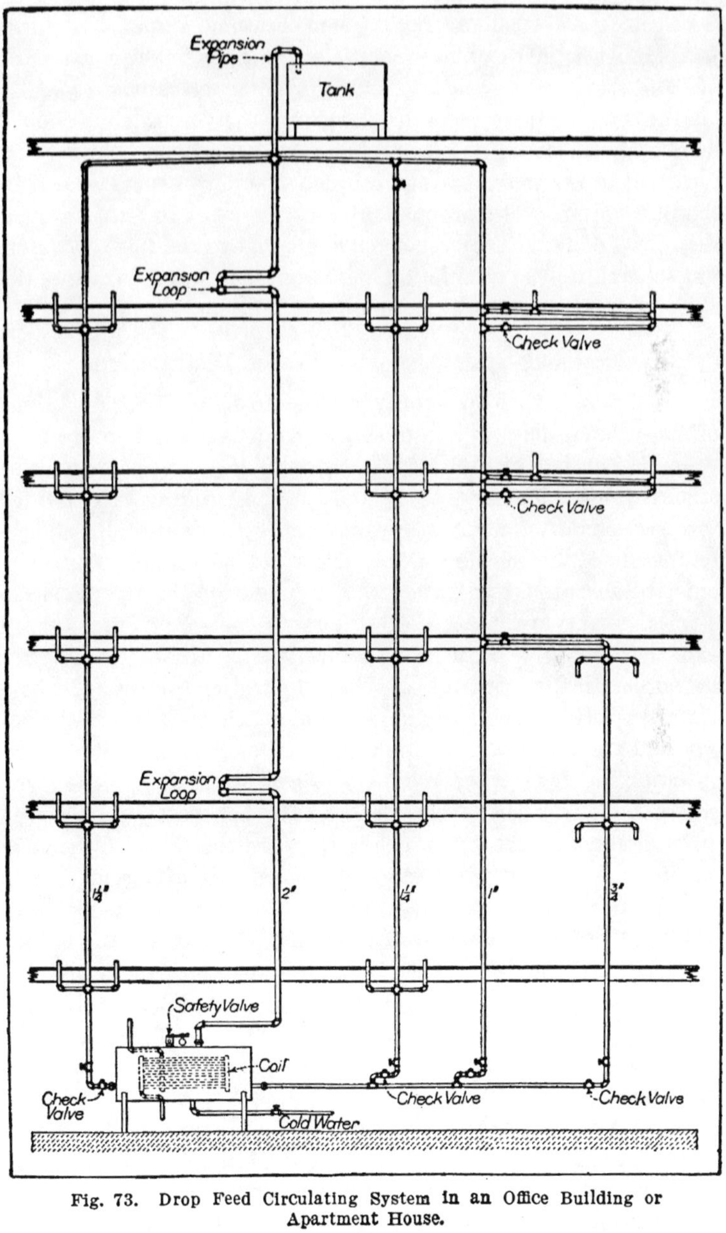

Fig. 73. Drop Feed Circulating System in an Office Building or Apartment House.

and this enters the tank through a check valve as in the case of the drop feed system and for the same reason. Expansion swing joints must be made on each of the rising lines in the same manner and it is well to provide for this also on the falling pipe. A lateral connection to some distance from the vertical line should have a circulation pipe also and if more convenient this may be returned to the basement as an independent return as shown in the illustration. The proportioning of the pipes in either system should be done with care not only to equalize the flow of water, but to equalize the circulation and so provide water of about the same temperature at all points of the building.

Sectional System of Hot Water Distribution.

Another system of supply which can be used for buildings of exceptional height or in places where it is desired to limit the pressure carried in the lines is shown in Fig. 75. This may be termed the sectional supply system, as the buildings are divided into sections of as many stories as may be desired and each system is in reality a separate one. To avoid the excessive pressure entailed in supplying water from a height of twenty or thirty stories, tanks are placed at different levels and the supply can be pumped either periodically to these or it can be automatically controlled. The illustration shows such cold water supply tanks placed side by side with the hot water heating and storage tanks. In this case the tank for the cold supply is supplying the hot water tank and cold water for all the floors in its section on the stories below. The hot water tank is supplying water to the floors above it up to the level of the next section. The illustration shows two methods of circulation in operation, one of them being the drop feed system already illustrated and the other being a rising supply system. In this the branches for the fixtures are taken off the main in its upward path, and the vent pipe is taken from the highest point and returned over the top of the cold water supply tank. The return pipe then returns directly to the hot water storage tank, entering through a check valve to prevent water being drawn back by the flow at the fixtures. The pumps on the cold water supply lines in such an installation may be automatically controlled so that as soon as the water reaches the desired level in the tanks the current is switched off. If cylindrical storage tanks are

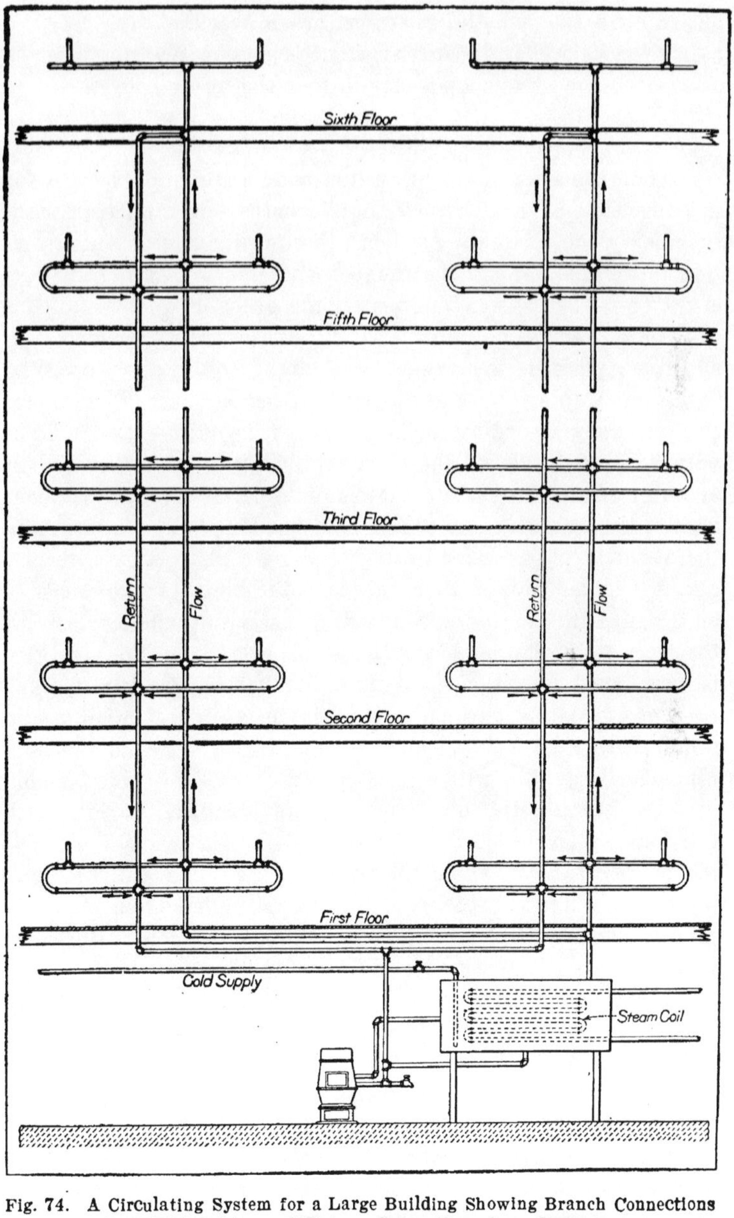

Fig. 74. A Circulating System for a Large Building Showing Branch Connections from Rising Pipe and Return Connections from Fixtures.

used a vent pipe may be carried to a point above a fixture connected with the waste pipe system or an overflow pipe may act as a vent as well to ensure that no excessive pressure will be carried on the system by the failure of the pump control.

Hot Water Supply to Shower Baths.

The proportioning of hot water heaters and supply pipes for shower baths in gymnasiums, public baths and similar institutions is a problem requiring more than ordinary care and attention to details. When the number and position of the showers is known it is necessary to estimate the probable quantity of hot water required to maintain a supply at the showers which will, when mixed with the proper amount of cold water, give the bather an ample supply at the desired temperature. The quantity will vary according to the design of the shower used. Thus with a 5 in. shower of the ordinary overhead type the amount of water required to give a shower of sufficient strength is about 4 gal. per minute and as the diameter of the shower is increased the quantity passed rises until it reaches 8 gal. per minute.

A heavier shower than this is not desirable, and this may be taken as the maximum that will be used by the bather. With the later type of needle, spray and shower baths, this quantity is very much increased. On the other hand there is not much need of estimating on a supply of water capable of maintaining a flow at each of the various shower and spray attachments simultaneously, as they will seldom or never be used in this fashion.

To estimate the quantity of water required to mix with the cold water to secure the ultimate temperature desired take the temperature at which the water in the boiler will be maintained and that at which the water is supplied from the main pipes. Then multiply the number of pounds of cold water to be raised in temperature by the difference between its temperature and the desired final temperature of the mixture and then divide that product by the difference between the final temperature of the mixture and the temperature of the hot water. The quotient will be the number of pounds of hot water that will be required to mix with the cold water to produce the desired temperature.

As an example, suppose that we have 1 lb. of water at 60 deg. F. and we desire to bring it up to 85 deg. F. by adding a certain quantity of water at 160 deg. F. How much will be

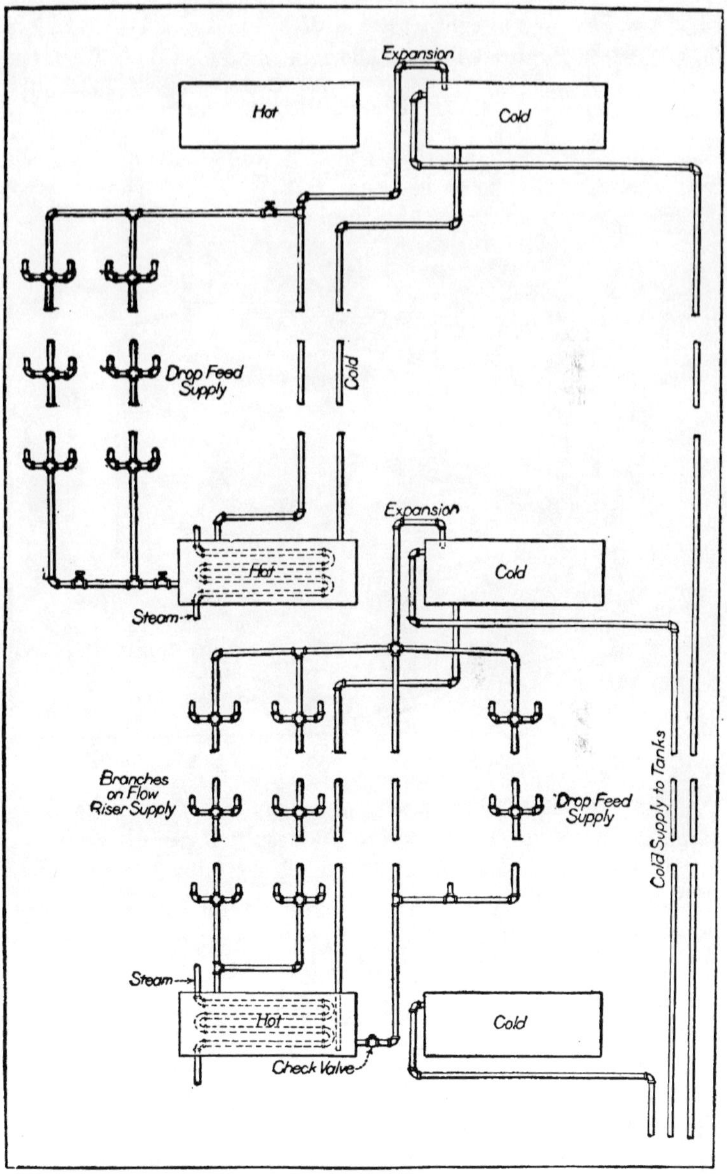

Fig. 75. Conventional Illustration of Sectional System of Hot Water Supply to a Very High Building.

required? Applying the rule, we have $85 - 60 = 25$ and $25 \times 1 = 25$. We also have $160 - 85 = 75$, and $25 \div 75 = 1/3$ lb. hot water required to heat the mixture to 85 deg. F. Or it may be stated like this: Quantity required $= \frac{1\ (85-60)}{160-85}$. This when further reduced becomes $\frac{25}{75}$, which equals 1/3 of the total quantity that is to be used.

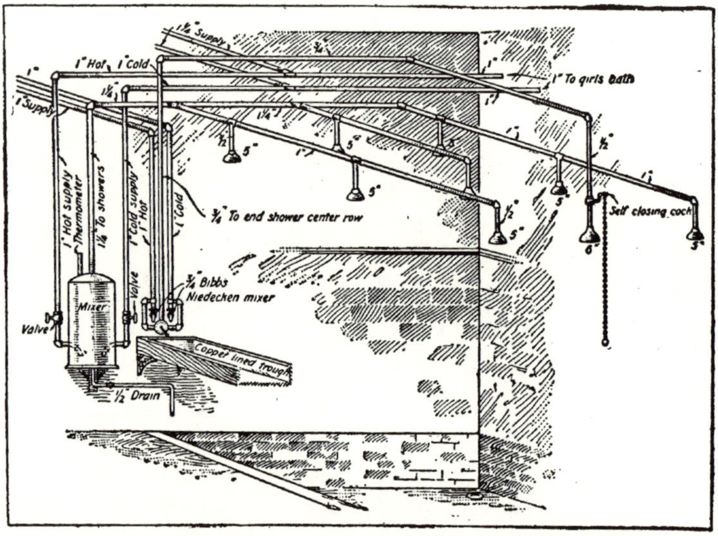

Fig. 76. An Installation of Showers in a Children's Home.

Now, to check up the accuracy of the foregoing, we will reverse the problem and work for the resultant temperature like this: One pound of water at 60 deg. F. contains approximately 60 B.t.u. One-third of one pound of water at 160 deg. F. contains approximately 53 1/3 B.t.u. and $60 + 53\ 1/3 = 113\ 1/3$ B.t.u., and 1 lb. + 1/3 lb. = 1 1/3 lb. total weight of water. Therefore, $\frac{113.333+}{1.333} = 85$ deg. F., or the required temperature.

When the amount of hot and cold water that is necessary has been found it is a comparatively simple matter to proportion the pipes according to the head of water carried so that the supply will be equalized and a sufficient quantity be available at each shower. The size of the heating boiler and coil or tank heater should be ample and should be proportioned on the maximum requirements of the installation. The method of estimating the size of coils and heating surfaces in boilers is explained elsewhere and all that will be necessary apart from that will be the exercise of judgment in the proportioning of the size of the storage tank. This will depend entirely upon how long the baths are to be in operation, how long each user is allowed to remain under the shower and how many bathers are to be provided for. The conditions in each case should be carefully considered and allowance made according to the demands that are to be made on the apparatus.

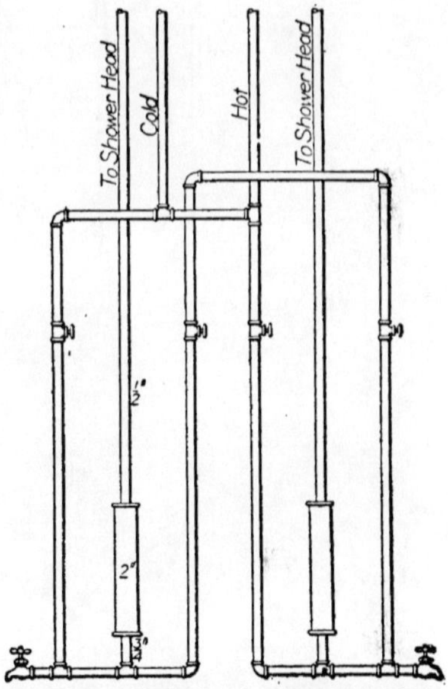

Fig. 77. Method of Connecting Hot and Cold Supply Pipes to Shower Baths.

Anti-Scalding Valves and Water Mixers.

Specially designed mixing valves are available for attachment to the supply pipes of shower baths, which will prevent any danger of scalding, as their construction and only possible means of operation makes it necessary to allow cold water to flow before any hot water is available. Thus the temperature can be adjusted to a nicety by the bather or attendant, but in some installations it is desirable to have the mixing chamber under the control of an attendant only, and in this case it may be installed as shown in Fig. 76. This shows a tank to which the hot and cold water is brought and into the body of which the water is introduced by bent pipes so that the two streams will mix and provide an equable temperature, under the control of the attendant, to all of the water flowing to the showers on the section supplied from the mixer. A thermometer fitted in the side of the mixing chamber enables the attendant to regulate the valves to supply the showers at whatever temperature he decides is required. For shower baths in factory washrooms and other places where an elaborate equipment is neither necessary or desirable, mixing chambers that will give satisfactory service may be constructed of pipe fittings as shown in Figs. 77 and 78. The

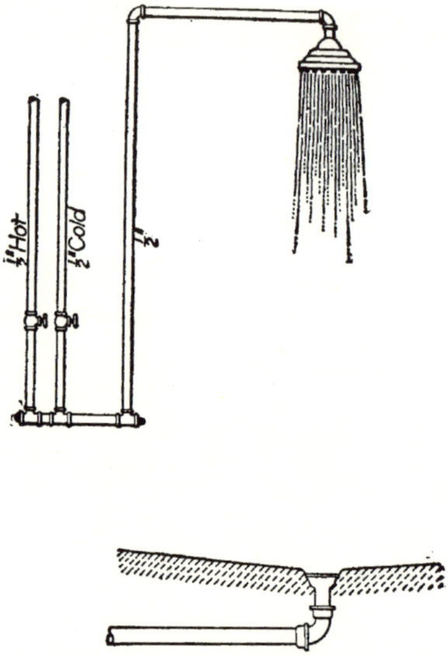

Fig. 78. A Simple Shower Mixing Pipe for a Washroom.

elaborate douche, shower and spray baths that are the features of spas, cures and some sanitariums are appliances that require special experience and knowledge of the purpose for which they are designed to successfully install and information that will be required to proportion systems in which such appliances will be used is best secured from the makers of the appliances.

Double Boilers, Connections and Distributing Pipes.

Where the water supply will not rise to fixtures on an upper floor, a tank is generally used to supply them, and in order to supply these upper fixtures with hot water a double boiler may be used and supplied from the same tank. The double boiler is made in different forms, both vertical and horizontal. Sometimes one boiler inside of another, and again two short boilers butting together, each connected with a separate water supply and sometimes with a special water heating device or receiver. The one in more general use is a boiler of smaller diameter inside of one of larger diameter. The outer boiler is supplied from the regular water supply and connected direct with the water back. The inner boiler is heated by the hot water in the outer boiler surrounding it and is supplied from the tank above the fixtures.

The same principles govern the operation and circulation of such boilers as govern in the ordinary single boiler. The receiver mentioned is made with two separate chambers so arranged as to secure an indirect passage of considerable length through which the water flows. One chamber is connected with both the water back and one of the boilers, and the passage of the heated water through it to the boiler heats the water in the other chamber, which is connected with the other boiler only. A more recent practice is to cast the water back with a division, making two separate parts and four openings, connecting a separate boiler supplied from the tank with one part in the usual way and another boiler supplied from the street service with the other part. This avoids cooling the water in the tank boiler when a large quantity of water is drawn from and enters the street boiler, which is experienced with the use of double boilers. The piping between the two boilers is so connected that both are sure to be supplied in case either source of supply fails.

Although double boiler work is a system of years' standing, the plumbers in general seem not much to blame for their lack of knowledge concerning it, as the conditions favorable to the use of the same can be found in comparatively few places. The plumbers who look no further than their present employment do not care enough to investigate, since they can make no immediate use of the knowledge. However, the truly ambitious plumbers are not satisfied until they are familiar with everything pertaining to their business, because they cannot tell how soon circumstances will place them where they will sadly need the information which at present is not required.

When the plumber is called upon to do a first-class job, it is often optional with him whether he puts in one or another kind of pipe. If, according to his knowledge, he thinks brass pipe will answer best, then brass pipe is used; but it is quite different in regard to the system to be employed. It is not so much a matter of choice as to whether the single or double system will be used or not. The proper conditions must exist before the double system can sensibly be preferred. A double system could be placed under almost any conditions, but such work in the wrong place would entail more work than would be necessary to place a double system in the right place, in addition to the difference in the original cost of the two systems.

Fig. 79 shows a double boiler system. Let us suppose that the street pressure will force the water into the tank in attic through A, instead of only to the second floor ceiling, for then the pump in the basement would be unnecessary. The inside boiler *a*1 and its system of pipe would also be useless. The pipe M could be continued to the fourth floor for cold water and branches made into J for hot water. If the street main furnished regular pressure and clear water, the tank could in some cases be omitted; but where the tank is omitted the auxiliary to constant pressure is lost— *i.e.,* where tanks are used settled or filtered water and regular pressure are assured, even though the street supply be shut off for repairs for hours, which is not unlikely. Were the pipe A delivering water to the tank from the street pressure, it would have to be furnished with a ball cock or something equivalent, instead of the bend at the tank, as shown. Any one can see the folly of using such a system as illustrated if the street pressure would reach the attic.

Where the conditions call for double system work, the plumber is called upon to select and adapt the style most suitable for the place. It will be understood that there are different

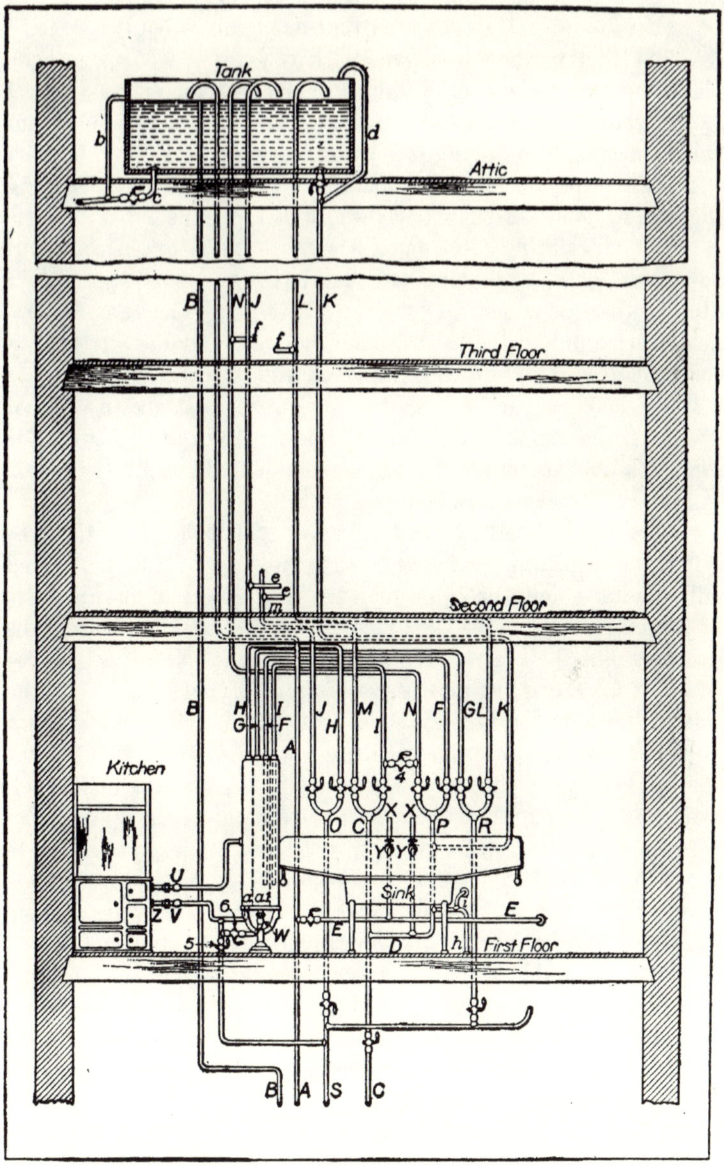

Fig. 79. Piping for a Double Boiler System of Hot Water Supply.

ways of arranging double boilers and the pipes leading to and from them, and yet give results that are practically. the same.

The first method used where there is available space is to place the two boilers independent of each other, either vertical or horizontal, as is most convenient. Having two independent boilers necessitates having two water backs—that is, one fire box with two water backs and connections from each back, making the circulation to each boiler independent of the other.

The circulating pipes must always be from one back to one boiler, unless a range with two fire boxes and two water backs each is used, in which case the tank pressure boiler may be connected with one water back in each fire box and the street pressure boiler connected to the two remaining fire boxes in the same manner. When such a range and connections are used, hot water can be supplied to both systems from either fire box. For some reason the boilers placed independent of each other seem to give the greatest satisfaction.

The second method is the placing of one boiler within the other. The difference between the capacity of the inner and outer boiler should be equal or a little in excess of the capacity of the inner boiler. The strength of the material for both shells can be about the same, and should be sufficient to withstand the effect of a vacuum without injury when formed into a shell the size of the outer cylinder. Should the inner cylinder of such a boiler be emptied or syphoned while the pressure is on the outer shell no damage would be likely to ensue, because the inner shell would only be required to support the weight of the water from the street, increasing in pounds per square inch according to the vertical head of water, in addition to the atmospheric pressure. The inner shell being naturally stronger from its smaller diameter, and having no side couplings to vary the strain or resistance, it would withstand any probable test without injury. It will be understood that the high, or tank, pressure is always connected to the inner boiler. A different result might be expected were the high pressure connected to the outer boiler during such a test as was mentioned above. In combination boiler work the water back connections are always applied to the outer shell, as one or the other must be heated by conduction.

Although there are few, if any, cases where a combination boiler has been heated by circulation through the inner shell or through both simultaneously from two water backs, there is no reason why the latter could not be done successfully. The inner cylinder should be made of copper, because it absorbs heat quickly. The outer shell, if also made of copper, will secure uniform expansion and make a much more durable job.

One way of arranging the pipes leading to and from a combination boiler is to supply a tank situated in the attic or upper floor from the street pressure by means of a pump upon the first floor or in the basement. The supply to the inner cylinder is taken from the tank, and is also connected to the street pressure, by which, should the tank supply fail, the street supply will fill the inside cylinder through a check valve. The tank and inside cylinder supply hot and cold water to all the floors above those for which the street supply can be relied upon.

Another method is substantially the same as the first, except the additional convenience of being able to send hot or cold water from the tank system to any fixture supplied by the street pressure by means of certain connections and stops properly placed in the kitchen.

A third way of using the double boiler system is as the first, with the addition of what is known as reverse cocks to the branches supplying the fixtures on the lower floors from the street pressure. The reverse attachment referred to has six openings and four stop cocks. They are set as follows: Upon the upper street pressure floor hot and cold branches from both street and tank supplies are brought to some convenient place and carried up through a safe pan, in order that any leakage from the cocks may be taken care of. Both of the hot supplies are connected to one leg of the attachment and cold supplies to the other leg. A lever handle is connected to the attachment cocks is such a manner that it is only necessary to pull up the handle to change from street to tank pressure, or *vice versa*.

A fourth arrangement of the pipes is a combination of the stop cocks in the kitchen, mentioned in the second method, with the reverse attachment, the reverse cock being placed upon the third floor when there is only an intermittent supply from the street to the third floor. Intermittent supply in some localities is caused by excessive drawing at certain times during the day, which in some cases causes the second floor to be uncertain if the street pressure alone is depended upon. Automatic attachments can be bought from any stock house for uses mentioned above. The object of double plumbing and everything pertaining to it is to avoid the cost of unnecessary pumping, storage capacity, etc., to as great an extent as possible. The true perception of the conditions existing in any case is the greatest aid to rightly determining which of the methods is best for the place, as well as whether combination or independent boilers are most suitable.

Double system plumbing is principally used in three, four-story and attic and five-story buildings, and the neatest examples of it can be found in residences. In high city buildings where high pressure steam is used both for heating and lifting water, other means of overcoming the irregular supply difficulty are found. It should be remembered that double boiler work and duplicate plumbing are not the same, the latter being merely a separate supply to each fixture and in some cases both separate and duplicate supplies.

The illustration is an example of double plumbing, which differs from the first method described only by having the stop cock No. 4 connecting the cold supply of both boilers above the sink. Should the street supply fail in this case, it is only necessary to turn stop cock No. 4 to supply the hot and cold street pressure system from the tank. A reverse attachment can be placed upon the second floor by simply making connections to N L from *e e* through the reverse cock. The range used in this job is of the ordinary type—*i. e.,* one fire box and one water back. Circulation takes place between the outer boiler *a* and the water back Z through the pipes U V. The emptying pipe shown by W is from the inside boiler *a*1. Its stop cock No. 6 is connected on the pressure side of cock No. 5, which prevents any possibility of the inner cylinder being emptied while the tank pressure is upon the outer cylinder. T is the sediment pipe through which both boilers must be emptied, and is controlled by stop cock No. 5. S is a general drain, which discharges over the basement sink. The pipes C and P have a small drain and stop to S from above the check valves, but are not shown in the drawing. Hot supplies are furnished with drains and cocks to S by continuations of O and R. The sink in this job is of porcelain, supported by legs and furnished with two drainers and with splash back. The drainers are supported by brackets, and the splash back can be removed by unscrewing the sink faucets Y Y and removing two wood screws at each end. The sink waste is indicated by *h* and the crown vent of its trap by *i*. The telltale pipe B discharges above the basement sink, that the person pumping may know when the tank is full. A is the supply to the tank in the attic, from a force pump in the basement. The pump suction pipe is connected to the street supply C. Tank drain *c* is

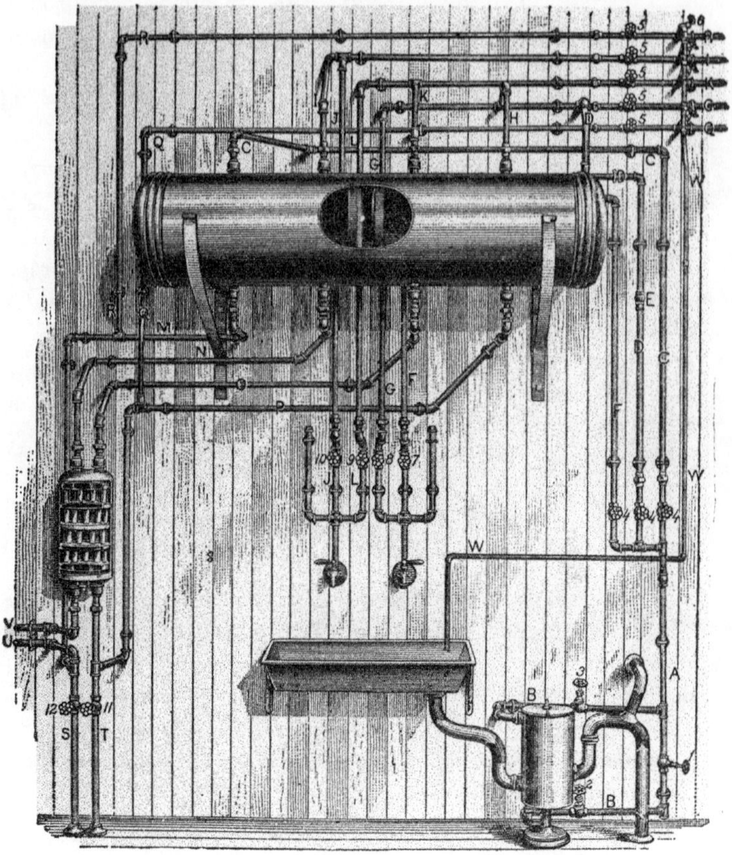

Fig. 80. A Horizontal Double Boiler and Connections.

furnished with a cock near the tank. The tank overflow *b* is connected to tank drain *c*. The tank cold main supply is first brought into the kitchen through K, thence through branch N to third and fourth floors, and up over the tank as shown, which insures the main line draining out should the water be shut off.

The inner cylinder is supplied with cold water through the branch F from K. Pipe *d* is branched into K below the stop cock as shown, which introduces the atmospheric pressure to the upper end of K, allowing K to be drained without draining the tank, should it be necessary to do so. The street pressure main cold is introduced through C and to the outer cylinder through branch I. Second floor cold is supplied from the street through branch pipe M. The kitchen sink, pantry sink and laundry hot water is supplied through pipes O, E, and their branches. Cold water to kitchen sink, etc., is supplied by branches from street pressure pipe C. Should the tank pressure fail, the street pressure will supply the inner boiler through branch D and a check valve; thence via P and F. The check is used to prevent mixing the tank and street water. Were a check omitted, high pressure would always be upon the outer boiler and all the water used would have to be pumped, by reason of the excessive pressure holding the check on pipe C shut. The check is placed upon street main cold C to prevent wasting the tank water into the street main when both systems are doing duty under high pressure; that is, when cock No. 4 is turned on. The check is also necessary to prevent drawing water from the outer boiler when the pump is in use. H is the main hot supply from the outer boiler, J being the distributing hot to second floor. G is the main hot from the inner boiler, L being the distributing hot to the third and fourth floors. Both L and J continue up to and bend over the tank in order to relieve any steam, vapor or expansion that may occur. X X indicate the air chambers from the sink faucets. It will be noticed that all pipes connecting to the top of the boilers are brought down to a convenient point above the sink to avoid using a stepladder when it is necessary to turn the stop cocks. The bends made in the hot pipe for the above reason prevent the successful use of a return circulating pipe. Both inner and outer boilers may have return circulation when the hot mains continue to rise above the boilers. The stops in this job above the sink are all plain stops. All jobs of the order here described should have the stop cocks and valves marked, and a chart giving full information as to the use of each one, both for regular service and in cases of emergency.

Horizontal Double Boilers and Connections.

A good example of piping work in the connections of a horizontal double boiler is shown in Fig. 80. The water back is not shown, but the pipes leading from it appear at the left, connecting with the receiver. The boiler is suspended over a kitchen sink that is connected with a grease trap. The apparatus consists of two separate boilers, butting together, one supplied from the street main and the other from a tank. The receiver has two chambers, one heated by being surrounded by the hot water in the other, which connects directly with the water back. AA is the cold water supply from the street main. BB is a by pass running cold water through the outer chamber of a grease trap to cool and harden the grease that collects on top of the water discharged from the sink. C is a branch to supply street boiler. D is a branch connecting tank and street supply to fill either boiler. E is a check valve to prevent tank supply from leaking into street when the latter supply fails. F is street supply to sink. G is tank supply to sink and by branch H to tank boiler and by branch D to street boiler. I is hot service from street boiler. J is branch from I to sink. K is hot service from tank boiler. L is branch from K to sink. M is cold water from street boiler to receiver. N is hot water from receiver to street boiler. O is hot water from receiver to tank boiler. P is cold water from tank boiler to receiver. Q is return circulating pipe from tank hot service. R is return circulating pipe from street hot service. S, sediment pipe from street boiler. T, sediment pipe from tank boiler. U, cold water from receiver to water back. V, hot water from water back to receiver. W, waste pipe to sink from cocks to empty upper pipes and fixtures when supply is shut off. The stop cocks are numbered. 1, stop to street supply; 2 and 3, stops to by-pass and grease trap; 4, 4, 4, stops to sink and street and tank boiler; 5, 5, stops to upper floor fixtures; 6, 6, waste cocks to drain pipes when upper floor fixtures

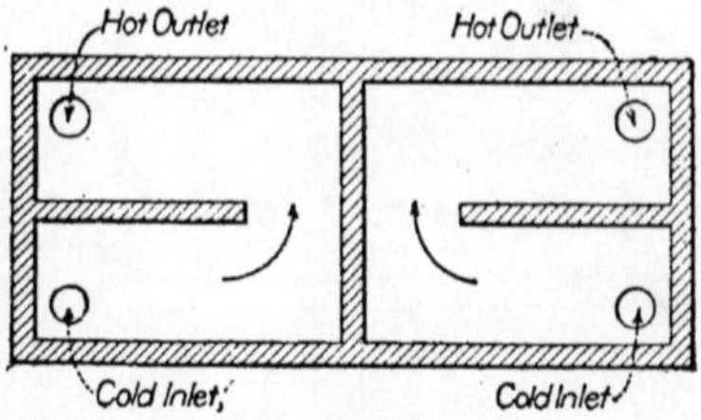

Fig. 81. Section Through a Double Water Back.

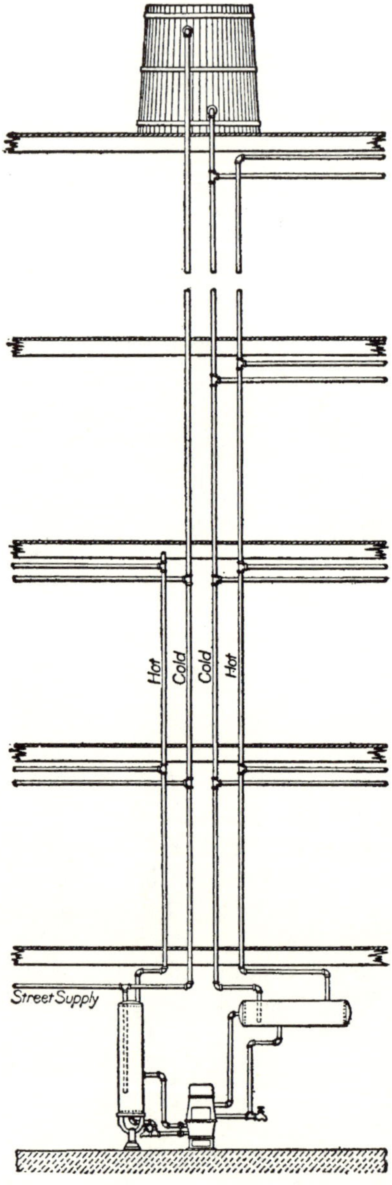

Fig. 82.—Method of Piping a Residence from a Double Section Water Heater.

are shut off; 7, stop to street supply when tank supply is used; 8, stop to tank supply when street supply is used; 9, stop to tank hot service to sink when street hot service is used; 10, stop to street hot service when tank hot service is used; 11 and 12, sediment stops to clean boilers and receiver. When tank is empty, tank boiler and kitchen fixtures can be supplied from street by opening stop cocks 4 on pipe D, 7, 8, 9 and 10. When street supply fails, street boiler and kitchen fixtures can be supplied from tank by opening stop cocks 7, 8, 9 and 10 and closing stop cocks 4, 4, 4.

The Use of Double Water Backs and Sectional Heaters.

A more or less common custom in cases where the pressure of water in the street mains is insufficient to raise the water to the upper floors of a building, but where it is desired to supply at least a portion of it directly from the mains and so avoid considerable pumping is to use two boilers. These are connected to a sectional water heater of which Fig. 81 shows the interior construction. This really amounts to two separate water heating systems heated by one fire, as one side connects to a boiler which is supplied directly from the city main pipe, while the supply to the other comes from the house tank on the roof. All the fixtures on the lower floors are supplied from the one that is connected with the street supply pipe, while the upper floors are connected with the house tank supplied boiler. A by-pass between the two may be made so that in the event of the water supply being short or being for any reason temporarily shut off, the lower floors may also be supplied from the house tank. In case a by-pass is fitted it is well to have a check valve on the street connection as there is less chance of the water being emptied back into the street main through the valve being inadvertently left open when this is fitted. It must be remembered, however, that it is necessary when a check valve is fitted to provide a safety valve on the boiler to take care of the expansion of the water in the system when heated. The method of making the connections to the different floors is shown in Fig. 82, which also shows a heater of two separate sections in use as an alternative to the double water back type illustrated in Fig. 81.

CHAPTER X.

Heating Water by Gas.

The value of illuminating gas as a means for heating water for domestic purposes is too well known to require any elucidation, but the manner in which the efficiency of any gas water heater may be ascertained is worth some attention. It is common to calculate the heating value of ordinary illuminating gas as being equal to 650 B.t.u. per cubic ft. In this case each cubic foot of gas consumed will heat 650 lbs. of water 1 deg. Fahr., as 1 heat unit is equal to raising 1 lb. of water 1 deg. Therefore the amount of gas that must be consumed to raise a given quantity of water to a desired temperature is easily calculated, and when the actual results are ascertained by noting the gas consumption, the percentage of efficiency of the heater is easily found. If, for instance, it is desired to raise the contents of a 250 gal. storage tank to 155 deg. and the temperature at which the water enters the tank is 40 deg., it will be seen that the difference is 115 deg. and that to raise each 1 lb. of water to that temperature 115 B.t.u. will be required. Multiplying 250 by 8.3 gives the weight of the contents in pounds, each gallon of water weighing approximately 8.3 lbs. Multiplying this sum by the difference in temperature gives the total B.t.u. required to heat the water. Then to find the number of cubic ft. of gas required divide by 650. In this case there would be 238,625 B.t.u. required, and this divided by 650 would show a requirement of about 367 cu. ft. Making allowance for loss of heat in flues and by radiation, probably about 500 cu. ft. would be consumed, showing a percentage of efficiency of about 70.

Efficiency of Gas Water Heaters.

Relative to the efficiency of the various types of water heaters, there are two kinds of efficiency: First, what is termed the initial test, made when heater is new and burners and heating surfaces are clean and conditions entirely favorable; and second, service efficiency, or the results obtained in practical use under average conditions, with the heater receiving only the normal amount of attention as to cleaning of burners and heating surfaces and flue pipes.

One of the most important things that lead to high efficiency in a gas heater is the vent pipe. If this has a good draft the burners will remain clean much longer because of the better combustion of the gas and there will be a corresponding increase in the amount of heat transferred from the fuel to the water. The efficiency of a gas heater will be affected as much as 20 per cent. by the good or bad draft in the vent pipe.

Fig. 83. Typical Gas Bath Heater.

The automatic - instantaneous heater of the better makes shows an initial efficiency of from 80 to 85 per cent., which is very high, especially when compared with heating and steam-generating apparatus in general. In the automatic-instantaneous heaters the service efficiency is also very high, usually showing from 75 to 80 per cent., even after the heater has had several years of service with only the average maintenance attention.

In the circulating tank heaters, the better grade of double copper coil heaters shows from 65 to 70 per cent. efficiency on initial tests; and, in service efficiency after years of use with only normal attention, this type of heater usually shows within 4 to 5 per cent. of its initial efficiency test. The cast iron tank water heaters of the better types show practically the same initial efficiency as the copper coil heaters, but service efficiency is not nearly as good and the maintenance cost is usually much higher than with the copper coil type.

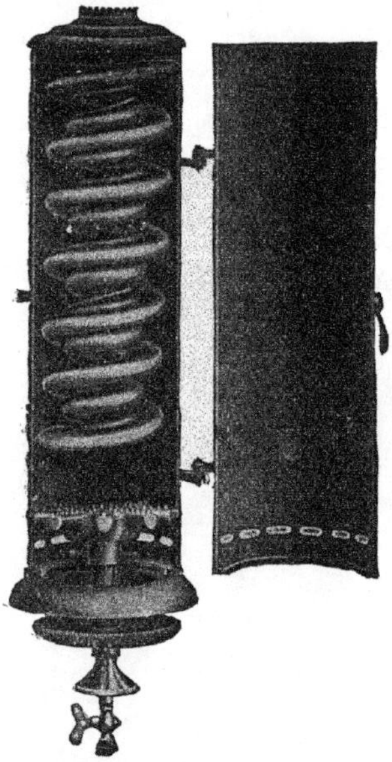

Fig. 84. Gas Heater of the Kitchen Boiler Type.

Lower Efficiency of Automatic Storage System.

In the automatic storage system the initial efficiency and service efficiency are less than with the automatic-instantaneous heater. It requires about 1¼ cu. ft. of 650 B.t.u. gas to raise one gallon of water 63 deg. F. on initial test, as against 1 cu. ft. for the automatic-instantaneous heater. On the service efficiency the automatic storage system does not hold up as well as the automatic-instantaneous, due principally to the fact that the continuous hard service on the burners causes them to become corroded or fouled very much sooner, and even with occasional cleaning and with tank heavily insulated it will usually require about 1½ cu. ft. of gas to raise one gallon of water 63 deg. F. with gas of 650 B.t.u. quality.

Constructive Features of Gas Water Heaters.

It is now pretty generally conceded that the copper-coil type of gas water heaters is preferable to all others. The very high heat conductivity of copper, combined with its high mechanical strength, makes it an ideal material for a water-heating receptacle. Being practically free from all tendency to corrosion, the water is never contaminated, the heating surfaces are easily cleaned, and viewed from all standpoints heaters of this construction far excel in durability and efficiency.

Removable and accessible burners of high heating power are desirable. They must be of simple construction and easy to clean and repair. Interchangeable mixer nozzles are demanded. In automatic heaters having thermostatic control, burners having copper gauze plates are preferred; whereas in the burners for circulating tank heaters the flame check screens are not favored, but the straight-drilled burner is preferred.

On the tank heaters the solid type of jacket is giving way to the open-front hinge-door type, with the pilot light eliminated, thus necessitating the opening of door in order to light the gas at the main burner.

In the automatic heaters the form of mechanism combining thermostatic or temperature regulation with water pressure control is the type generally favored by manufacturers of this class of water heater for general use.

A difference of opinion exists as to whether the internal or external type of thermostat is preferred, the internal thermostats being preferred by some on account of their being located nearer the seat of activity, while the external thermostats are often favored because of their circulating features and greater accessibility for adjustment and repair.

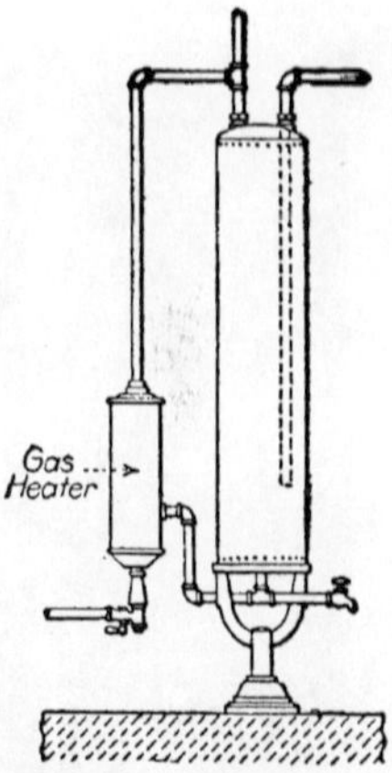

Fig. 85. Method of Installing Kitchen Boiler Heater.

The type of jacket for automatic heaters having the double cast iron wall with large air space between, and with doors of liberal size exposing the coil and burner compartments, has been adopted by all manufacturers, thus permitting easy cleaning of coils and burners or inspection while in use.

The automatic supplementary system of connecting heaters, whereby all water is first taken through the ordinary tank, heated by a coil in the coal furnace, is favored by practically all water heater manufacturers and used by practically all gas companies in making installations, especially in the cooler climates where buildings are commonly heated in this way.

Instantaneous Bath Heaters.

These appliances are of two general types which may be classed as contact and non-contact heaters. In the first, the water is in actual contact at one point of the heater with the heated products of combustion from the Bunsen burners. The second heats the water indirectly by spreading it over a large surface exposed to the gas flame.

This type of heater is generally of such attractive appearance that it is an ornament to the bath room or apartment in which it is placed.

The heater illustrated in Fig. 83 is constructed of sheet copper, nickel plated, with valves and supply pipes of brass, nickel plated and polished. The heater consists primarily of a number of Bunsen burners placed under a shell, over which water is caused to pass in a thin sheet by being sprayed over a perforated cone, through which it passes exposed to the intense heat from the Bunsen burners and so down over the lower part of the shell which is in the form of a frustum of a cone to the outlet tube.

Fig. 86. An Automatic Water Heater Attached to a Vertical Boiler.

When more than one fixture is to be served a gooseneck or offset is provided so that an open end can be maintained to safeguard the heater. A valve can be fitted to the other outlet so that water can be drawn freely, without the necessity of running any through the open end to the other fixtures. A by-pass or pilot light is arranged so that it must be turned on and lighted before the main gas valve is opened, thus insuring that gas will not collect in the casing before the light is applied. The water supply valve is also arranged so that the lever must be automatically turned up by the action of opening the gas valve.

The whole heater is placed on a white enameled iron shelf supported by a bracket. The Bunsen burners are readily removed for cleaning when necessary. A pipe must be connected from the top of the heater to some convenient flue or carried to the open air to remove the products of combustion. When taken to the open air a hood should be fitted so that down draughts will be prevented, but the vent should enter a heated flue whenever possible.

The second type, the simple kitchen boiler heater, is of a different construction to the type just described. Here the water is circulated as it is heated to a storage tank, from which it is drawn as required. The heater generally consists of a coil of copper pipe, as shown in Fig. 84, or a series of hollow discs connected by short pieces of pipe which presents a comparatively large heating surface to the effect of a hot Bunsen flame produced by a burner placed at the base. The heated products of combustion pass through the coils of pipe or over and around the hollow discs in their passage to the outlet at the top. This outlet must in all cases be connected with a heated flue or carried outside and finished with a hood for the same reason as that for which the bath heater was connected.

Fig. 87. Pressure Controlled Gas Water Heater.

The range boiler heater is generally connected as shown in Fig. 85, that is the flow connection is made to the upper tapping of boiler just where the house supply is taken off, while the lower pipe is connected at the bottom tapping of the boiler. Should the water be of such a nature that there is danger of sediment choking the boiler or coils the lower connection may be made to the side opening of the boiler. This, however, decreases the hot-water storage capacity by one-half, and a better plan is that where a sediment trap or other means of collecting the sediment before it passes into the heater is provided. Reducing the storage capacity, however, allows the water contained in the boiler to become hotter. The position of the coil in relation to the Bunsen burner in this type of heater is clearly shown in Fig. 84.

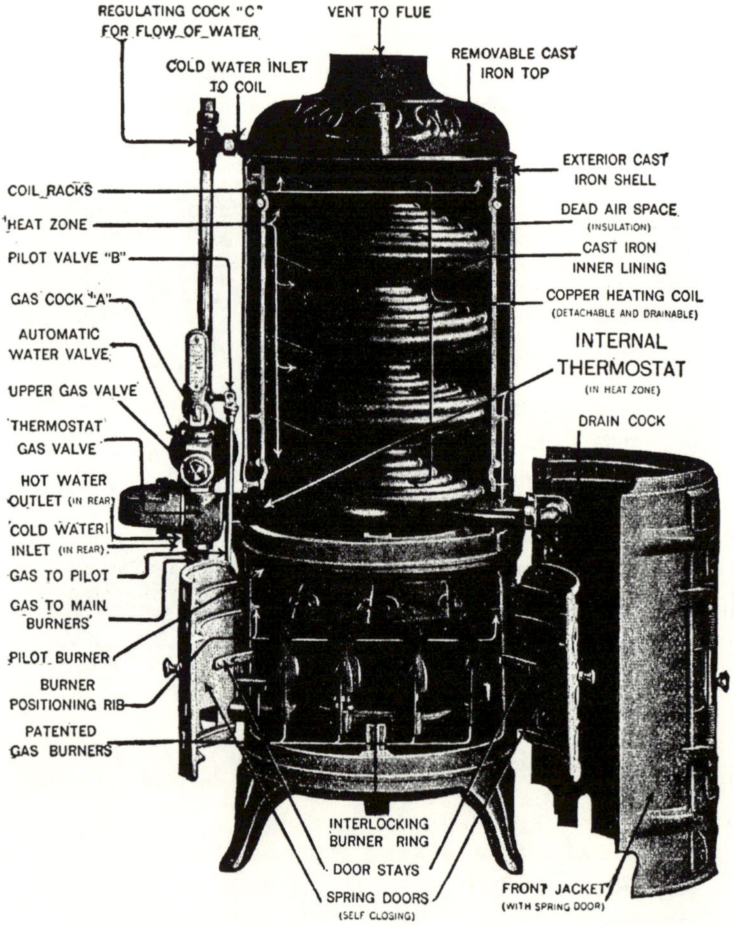

Fig. 88. Details of a Pressure and Thermostatically Controlled Heater.

When the hollow disc type of heating surface is used these take the place of the coil and occupy the same relative position, so that the heated gases pass over the surface in their path to the outlet at the top of the casing. The water connection itself is generally relied upon to support the heater, but brackets can be placed to support it when necessary.

The third type of gas heater in the list, that shown in Fig. 86, is simply a refinement of the type just described. This type is becoming increasingly popular for the reason that it is economical in fuel. The refinement consists in the provision of a gas valve thermostatically controlled. The thermostat consists generally of a rod of some material (porcelain is commonly used), which will expand to a considerable extent with the rise in temperature. The arrangement is such that the expansion will by a combination of compound levers operate the valve controlling the gas supply. Thus if it is desired to maintain the water in the storage tank at a temperature of 160 deg. the gas will be automatically turned on and ignited by a pilot light as soon as a fall in the temperature in the boiler great enough to cause the rod to contract and open the gas valve occurs. The valve will remain open just so long as the temperature remains below the predetermined point. When it has been reached the expansion of the porcelain rod is sufficient to close the valve and to hold it closed until the water cools again.

Fig. 89. Automatic Heater with Thermostat Placed on the Outside.

The fourth type is built along different lines, but also has the economical feature highly developed. This type of heater, which is shown in Fig. 87, is not intended for use with a storage tank but is intended to supply hot water instantaneously on the opening of the faucet. This is accomplished by means of a pressure valve which may be seen at the side of the heater immediately over the gas supply valve. The normal pressure in the water supply system, be it 10, 20 or 100 lb. per square inch, is caused to act on this in such a manner that when all the faucets on the system are closed it will maintain the plunger in such a position in its cylinder that the rod which connects it with the valve in the gas supply line is closed. As soon as the pressure on the coil side of the water valve is relieved by the opening of a faucet on any part of the line, this plunger moves and allows the gas valve to open, thus allowing gas to pass to the burner. The pilot light, which is contained in the burner chamber, immediately ignites the gas and the water is heated instantaneously as it passes through the coils to the faucet. Thus no gas is consumed except at such time as the faucet is open.

To get the maximum economy the heater should be as near the fixtures as is convenient and the pipes should be no larger than absolutely necessary to provide a steady flow of water with the pressure available. These valves are really very simple and are not likely to cause trouble. The connections are extremely simple. A cold supply to one side of the valve, a connection from the coil to the hot-water lines and a connection to the gas valve are all that are required.

This type of heater is being largely adopted for suburban residence use and as it does away with the necessity of storage tanks it is well suited for apartments where space is limited, more especially as it can be readily placed in a corner of the kitchen or in a closet if a flue for the discharge of the products of combustion is available. This heater has a capacity of 2 to 3 gal. of water raised 60 deg. above the temperature of the cold water per minute and its makers recommend that a pressure of water of at least 8 lbs. per square inch be available for operating the valves. The heater is illustrated in Fig. 87.

The next type includes the provision of both pressure valves and thermostatic valves, an arrangement which on the larger sizes is likely to effect a considerable saving of gas in this way. When the faucet is opened the gas is immediately ignited and heats the water to a certain temperature. If a thermostat is provided a definite temperature may be decided upon and immediately this is reached the thermostat will close the gas valve. The water still keeps passing through the coil and just as soon as the temperature falls below the point decided upon it will automatically open the gas valve again and keep it open until the temperature again rises above the required amount.

This balances the varying pressures of gas at different periods, as the first time the gas pressure is high more gas than would be necessary might be passing and the water would be unnecessarily hot. On the other hand for bathing purposes less

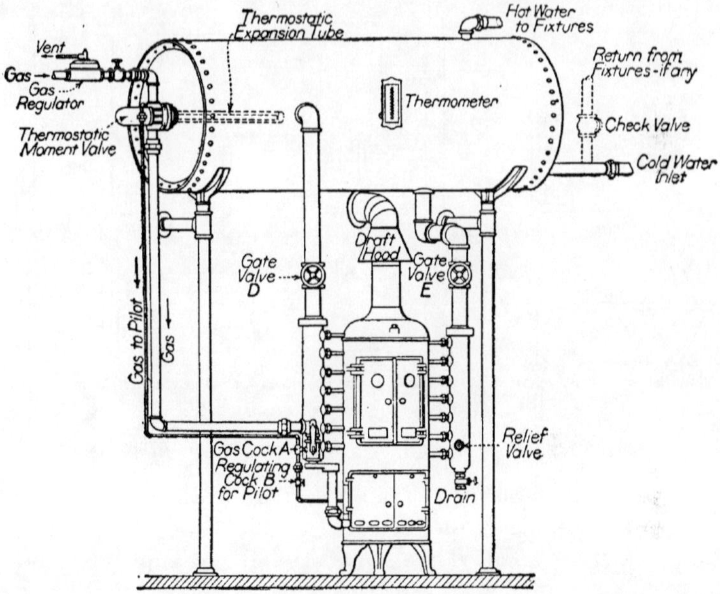

Fig. 90. A Heater of the Storage Type with Gas and Water Connections.

would be required, but when the fixtures are used by many members of the family this point is often lost sight of and much hot water is allowed to go to waste. The thermostat takes care of this and provides the same amount of water at the same temperature to all without special care or attention.

A heater of this type is illustrated in Fig. 88. The amount of pipe in the coils will be noticeable. The illustration clearly shows the position of the thermostatic and pressure valves and the pilot valve and tube. A heater of a similar type but having the thermostatic valve placed on the outside of the heater in a tube connected with the coil is illustrated in Fig 89.

For the larger size of heating installation, that is for institutions where large amounts of hot water are required, the last type, the automatic heater with thermostatic control but no pressure valve is the best adapted. This type is shown in Fig. 90. As will be seen it is practically the same as that shown in Fig. 88, but is of a much larger size. The thermostat may be placed either in the tank or the lower part of the heater.

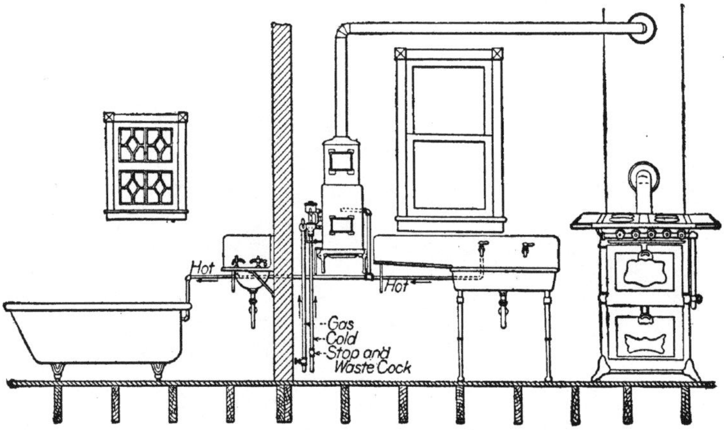

Fig. 91. Method of Installing Automatic Heater in a Flat.

In the first case it is shaped so that a casing is inserted through a tapping in the tank and in this casing the expansion rod is carried. The valves operating the gas supply are in a casing at the outlet end of the rod as shown in Fig. 90.

In the other case the rod is contained in a casing which is placed in the burner chamber of heater and connected so that the heated water acts directly upon the expansion rod. As it lengthens with the rise in temperature in the coil, it acts on toggles again and closes the gas valve at whatever point may have been decided upon. This thermostat is shown in Fig. 88, which also shows the Bunsen burners and the coils inside the casing.

The doors on most of these heaters are mounted with springs so that should gas be collected for any reason before being ignited by the pilot light, any excess pressure will be relieved by the yielding of the door. This obviates all risk of damage by the explosion.

The best methods of using and maintaining in good condition all of the various types of heaters are amply described in the various makers' catalogues and instruction sheets. These sheets are invariably sent out along with the heaters and should be carefully studied by those not familiar with the installation of

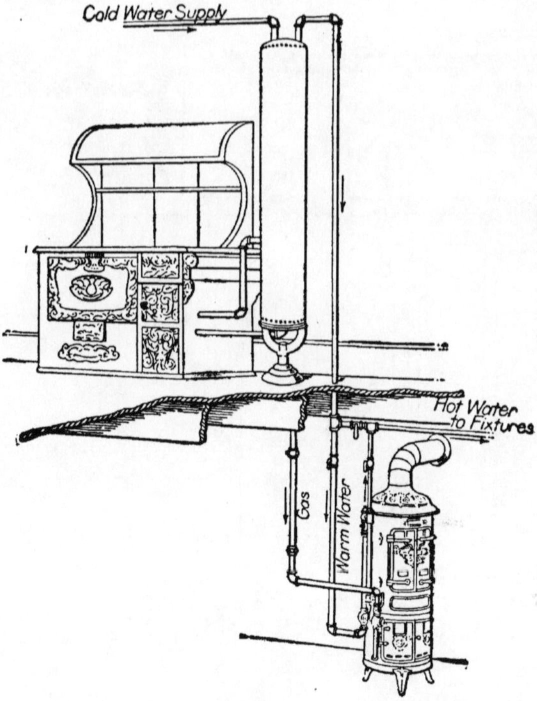

Fig. 92. A Heater Connected as an Auxiliary.

such machines and should also be brought to the attention of the customers so that mutual satisfaction will be obtained. For the guidance of those who have not previously installed gas water heaters a few rules may be of use.

1. Be certain that the pressure of water is such that the supply will always be available at the highest fixture on the line.

2. Do not use a heater of the automatic pressure valve type unless the water pressure is equal to the amount designated by the maker of the heater.

3. Be sure the gas supply is ample.

4. The vent pipe should never be reduced below the size of the outlet collar on the heater and should not have a damper.

5. Always flush out the supply pipes before connecting to the gas heater. This is important, as chips or lead used in making the joints may lodge in the valves and cause trouble.

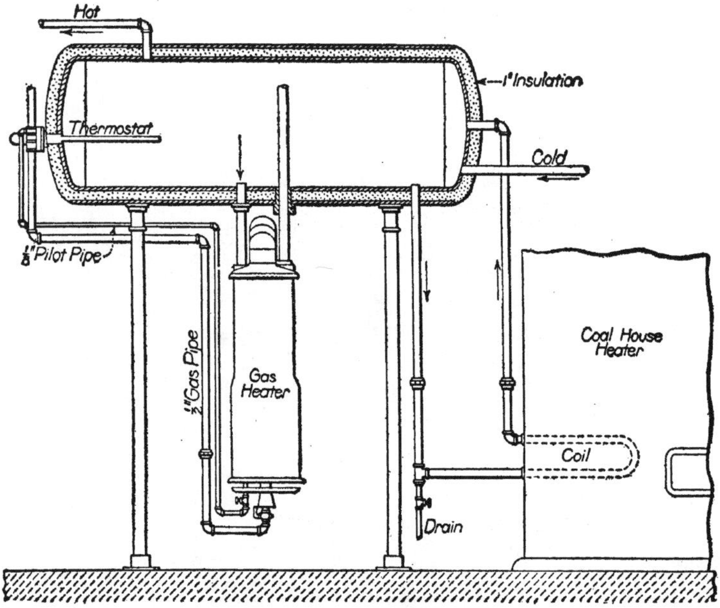

Fig. 93. Automatic Gas Heater and Coil in Furnace Connected to Large Storage Tank.

6. Check the supply to the automatic heater down to what the capacity of the heater calls for. This will obviate complaint of lack of heating power through the customer drawing water faster than it can be heated.

Various Methods of Connecting Heaters.

It has already been stated that ordinary boiler heaters can be connected up in such a way that sediment will not collect in the coils. Should a coil become choked it will manifest its condition by snapping sounds due to the formation of steam and the supply of hot water will be unsatisfactory. The coil should be removed and thoroughly cleaned and the connection made either with the return pipe connected to the side outlet of the boiler or connected through a sediment chamber which will collect it at the lowest point. Fig. 91 shows an automatic heater placed in an apartment and supplying kitchen and bath room fixtures, while Fig. 92 shows how they may be connected as auxiliaries so that the coal stove can be used at any time and hot water drawn from the boiler through a by-pass. If desired the storage type of heater may also be connected as an auxiliary to a coal stove and the thermostatic control may be used to maintain the water at any desired temperature at such times as it may be desirable to have a low fire in the coal stove.

These heaters may be connected in many different ways, but do not differ greatly from other heaters, and the plumber who understands the principles which cause water to circulate will have no difficulty in installing any or all of the heaters.

The method of connecting a coil in a house heating furnace to a boiler also heated by an automatic gas heater is shown in Fig. 93. The illustration also shows how the insulation of the tank is put on to effect economy in gas consumption.

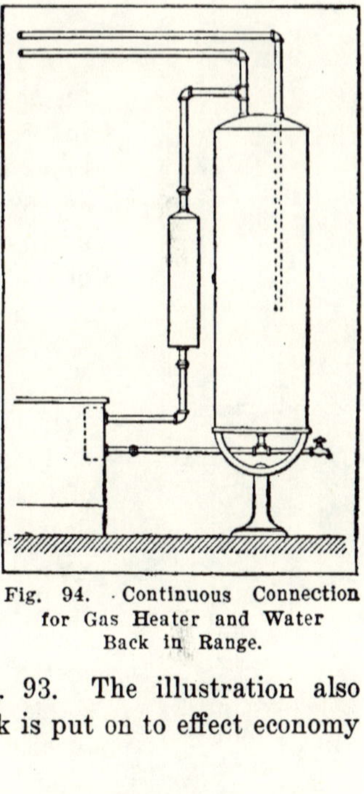

Fig. 94. Continuous Connection for Gas Heater and Water Back in Range.

Kitchen Range and Gas Heater Continuous Connection.

A continuous flow connection with a gas water heater of the common kitchen boiler type and the water front in the range is shown at Fig. 94. This is a neat method of connecting the two fixtures which saves space and also prevents to some extent short circuiting of cold water through one or other of the heaters when one is not being used. The coils of the gas heater will radiate a little heat when the hot water from the water front is passing through them, but not enough to offset the advantage possessed by the method in other respects. A double boiler connection may be used to connect the flow pipe with that of connection of the boiler so that the supply to the fixtures will be drawn from the boiler and there will be no chance of short circuiting it and drawing through the water front instead of from the boiler.

A Space Saving Method of Installing a Gas Heater.

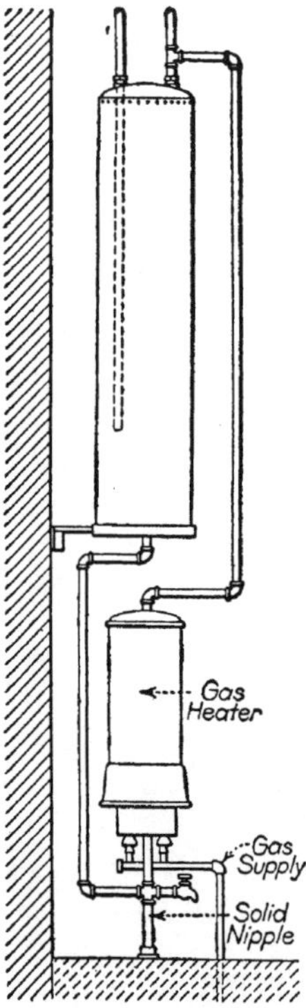

Fig. 95. A Space Saving Method of Installing a Gas Water Heater.

A departure from the usual method of fitting kitchen boiler water heaters is shown in Fig. 95. In this arrangement space is saved and a neat appearing job is secured as well while the method admits of connecting the flow pipe to the side connection of the boiler if desired. The usual practice of placing the same size boiler as is used with a coal range and water back is open to the objection that the expense of heating up the entire contents of the boiler is heavy and as the heater is, as a rule, capable of heating the water almost as fast as it is drawn, the amount of storage called for with the use of a water back is unnecessary. If then a 20 gal. boiler is placed in the manner shown and the connection is made to the upper tapping, the contents of the boiler can be heated to whatever temperature is desired, and if this is not a large enough quantity for the purpose intended, the gas can be lit and the water heated while it is being drawn, thus using the hot water already stored with the additional supply being heated. If the boiler has a thermostatic control valve attached this will bring the burner into action as soon as the water is drawn and so avoid the necessity of attention on the part of the user. The smaller boiler will also show some economy in gas, as there is less loss of heat by radiation and of course less water to maintain at the desired temperature.

The method of installing the heater under the boiler is especially well adapted for positions where the ordinary method

some awkward pipe fitting. Such a position, for ften seen in the closets behind kitchen ranges in two or threé apartment style where space is lime it is common to install the boiler in this closet. e well to state that when a gas heater is placed in a r position where it is in close contact with woods should be covered with tin to a sufficient distance e heater to prevent its ignition through overheatshould be nailed to cleats on the wall for the purding an air space behind it, and asbestos board on below the tin. It is also important to see that a nnection is made to carry off the burned gases and may leak through the burner accidentally.

CHAPTER XI.

Heating Water by Steam Coils and by Injecting Steam and by Coils in Heating Furnaces.

The designing and proportioning of heating systems for either domestic or industrial hot water supply on a large scale generally contemplates the use of steam as the heating medium. This may be either the sole means of heating or only an auxiliary, but in either case a considerable amount of care must be given to the proportioning as well as to the actual construction to secure satisfactory results. The first thing that is necessary is to find the amount of heat that will be required to maintain the quantity of water required at the desired temperature. Then the amount of steam that is required to convey that heat to the water and the pipe sizes that will convey the necessary amount of steam must be ascertained. There are quite a few factors that must be taken into consideration in arriving at the required data. The transmission of heat through the steam coil to the water varies according to the nature of the metal used, and if a close approximation to the actual steam consumption and pipe sizes is to be arrived at this point must receive careful study.

Data on Heating Water by Steam.

A French author is responsible for the statement that for the same difference of temperature between steam and water the coefficient of the flow of heat through the wall of a coil of pipe filled with steam and placed in a tank of water varies from 1 to 10, according to the speed of the circulation of the water. It is said that 1 square meter of copper tubing placed in a liquid in motion can condense about 3 kilograms of steam per sq. meter per hour per degree of difference in temperature between the steam and the water, which figure is equivalent to 0.34 lb. of steam per sq. ft. per deg. Fahr. per hour, or about 329 B.t.u. This figure corresponds to steam at atmospheric pressure and water entering at say 140 deg. and passing away at 194 deg.

With an iron pipe coil having 1 sq. meter of surface placed

in a tank holding about 265 gal. it was found that without circulation in the tank, that is, with the water left intact in the tank, the water was heated from an initial temperature of about 53.6 degrees Fahr. to the temperatures noted in the accompanying table. Opposite each temperature is given the number of minutes required to warm this volume of water from the initial

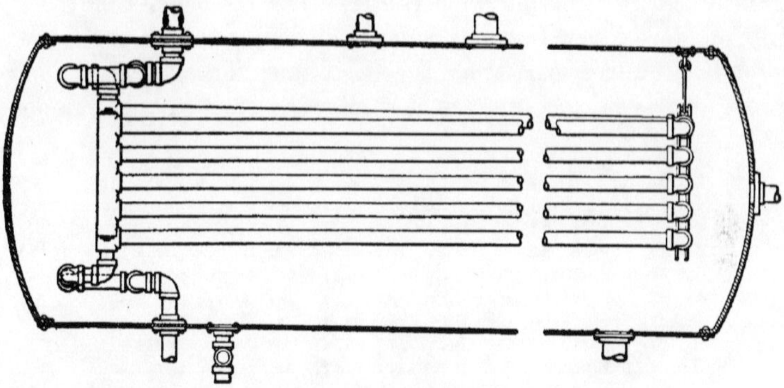

Fig. 96. Arrangement of Steam Coil in Hot Water Tank to Allow for Expansion.

temperature to that stated, both for low pressure and for high pressure steam:

From 53.6 degrees to	Steam at 1½ to 3 lb.	Steam at about 60 lb.
122 deg.	in 35 min.	in 33 min.
140	44	42
158	60	51
176	95	60
194	120	68
212	..	77

A test was also made of the heating of water passed continuously through the tank. Apparently the same steam pipe coil was used, presenting 1 sq. meter or 10.75 sq. ft. of surface. It was found that starting with the water at about 53.5 degrees, 1½ or 1.6 gal. were heated per minute to 176 degrees Fahr.; 2½ gal. to 167 degrees; 2.9 gal. to 158 degrees; 3.3 gal. to 149 degrees; 3.8 gal. to 140 degrees, and 4.75 gal. to 122 degrees. These figures were obtained with the steam at about ½ to 3 lb. pressure. With the same arrangement, but with the steam pressure at about 28 or 29 lb., about 9¼ gal. were heated to 140 degrees, as compared with 3.8 gal. with steam at low pressure.

The author quotes from data of Thomas and Laurens, who

give the transmission of heat through a copper tube through which the water is passed with the tube placed in a chamber of steam. The tube was 10 millimeters (0.39 in.) in diameter, with the copper 0.04 in. in thickness. The total length was about 10.3 ft., so that its surface amounted to 1.06 sq. ft. The steam was maintained at atmospheric pressure. The coefficients of transmission were given for a velocity of the water at 0.1 meter per second and for speeds two, three and four times as great, and so on. These figures indicate the following rates of transmission of heat in B. t. u.:

0.4 ft. per second, 335 B.t.u. per sq. ft. per degree per hour.
0.5 ft. per second, 390 B.t.u. per sq. ft. per degree per hour.
0.7 ft. per second, 455 B.t.u. per sq. ft. per degree per hour.
1 ft. per second, 510 B.t.u. per sq. ft. per degree per hour.
1.5 ft. per second, 560 B.t.u. per sq. ft. per degree per hour.
2 ft. per second, 610 B.t.u. per sq. ft. per degree per hour.
2.5 ft. per second, 655 B.t.u. per sq. ft. per degree per hour.
3 ft. per second, 700 B.t.u. per sq. ft. per degree per hour.
3.5 ft. per second, 750 B.t.u. per sq. ft. per degree per hour.

These figures are applicable to the ordinary type of water heater with a coil supplied by steam and the type commonly known and used as feed water heaters in which the water passes through the coil exposed to the heat of the steam in a jacket. This latter style is occasionally used for hot water supply in factory wash rooms and laundries where exhaust steam is available with satisfactory results, and the rate of transmission is higher with this type, as the foregoing table shows. In the ordinary type of horizontal tank with brass or copper coils it is common to allow about 1 linear ft. of 1 in. pipe to each 5 gal. capacity. This works out at about 1 sq. ft. of heating surface to each 15 gallons of water, and with steam at about 225 deg. temperature the contents of the boiler should be heated in about one hour to a temperature of about 170 deg. The proportions given in the French tests fall a little short of this, but the amount of heating surface necessary is best determined by ascertaining the probable requirements of the building and making allowance according to the pressure of the steam supply available. The coils usually fitted in horizontal boilers of the style used for hot water supply are built of brass or galvanized iron pipe with return bends or headers. In either case there should be provision made for expansion by making swing joints on the connections as shown in Fig. 96. This also shows the connections made through the the boiler shell in such a way that the coil can be removed from the boiler when necessary to replace a leaky pipe or fitting. Many engineers believe that better service is secured by the use of galvanized iron headers or return bends and ells rather than cast brass, owing to the liability to spreading of the fitting when making the connections. The forcing of the pipe into the last thread in the endeavor to secure a thoroughly tight joint will occasionally cause this to occur

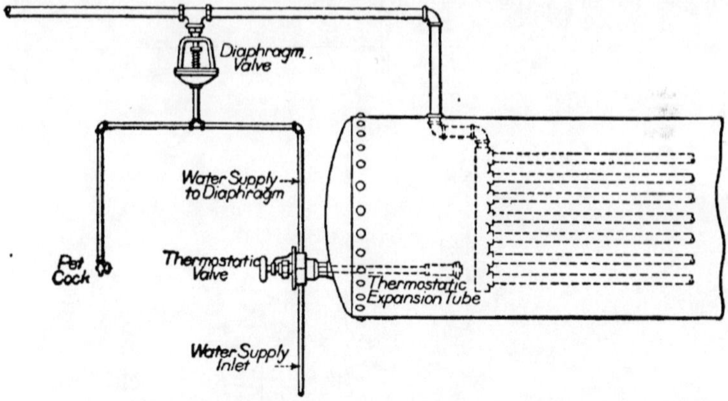

Fig. 97. Thermostatic Control of Steam Supply to Coil.

and where a high pressure is carried in the water supply the water will find its way into the steam heating system and cause trouble through raising the water level above its normal height. A pressure relief valve is usually fitted on the boiler and a thermostatic valve to control.the steam supply. This acts through the lengthening of a rod by expansion to close the steam valve or by the expansion of a liquid contained in a closed chamber which is caused to open and close the steam valve through a diaphragm movement. Fig. 97 illustrates a thermostatic valve operated by expansion. This opens a small valve on a pipe connected with the cold water supply and the pressure is then caused to operate on a diaphragm connected with a steam valve. Thus when the water in the boiler reaches the desired temperature the expansion of the rod opens the small valve and the pressure then closes the steam valve. As the temperature falls the thermostatic element in the tank contracts, allowing the valve on the steam line to open again and to remain open until the temperature rises again to the desired height.

Heating Kitchen Boilers by Steam.

Heating ordinary vertical kitchen boilers by steam is not so often practiced, although it is done occasionally. The coils in this case are generally spiral and made of copper or brass pipe. The steam enters at the top of the coil and the return pipe carries off the water of condensation. The usual allowance of

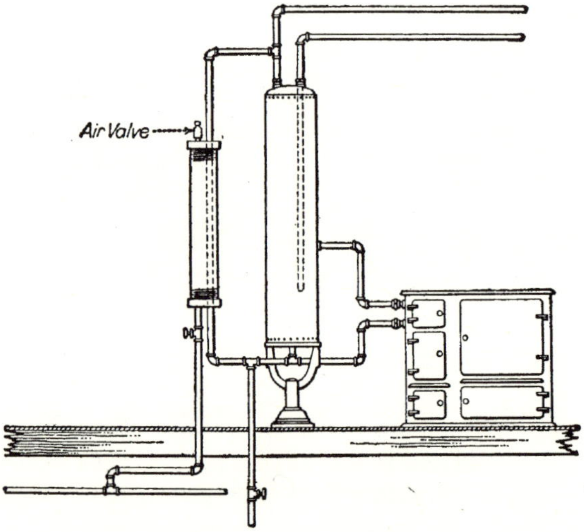

Fig. 98. Kitchen Boiler Heated by Steam.

1 linear ft. of 1 in. pipe to 5 gal. capacity would therefore call for a coil of 8 ft. of pipe in a 40 gallon boiler and this is easily provided. A simple steam heater which is fitted in the same manner as the ordinary kitchen boiler gas heater has also been used. This consists of a length of galvanized iron or brass pipe through which passes the cold water supply from the bottom of the boiler. Steam is introduced to the large pipe as shown in Fig. 98, and transmits its heat to the circulating pipe passing through it. The water as it is heated rises and passes into the boiler, being replaced with the cooler water from the boiler, and thus a circulation is maintained. The length of the pipe is determined by the size of the boiler and the time allowed

to heat the contents. It is somewhat difficult to allow as much heating surface as is done when the steam is passed through a coil, unless it is possible to continue the large pipe through the floor into the basement or room below the boiler but the rate of transmission of heat from the steam to the water inside the circulating pipe is a little higher than when the process is the reverse as with a coil in a boiler.

Heating Water by Injecting Steam.

In large institutions and in business places such as laundries and cleaning establishments it is sometimes found desirable to heat the water required for washing and other purposes by injecting live steam into tanks at a high pressure, the tanks being of course open, or at least not under any water pressure.

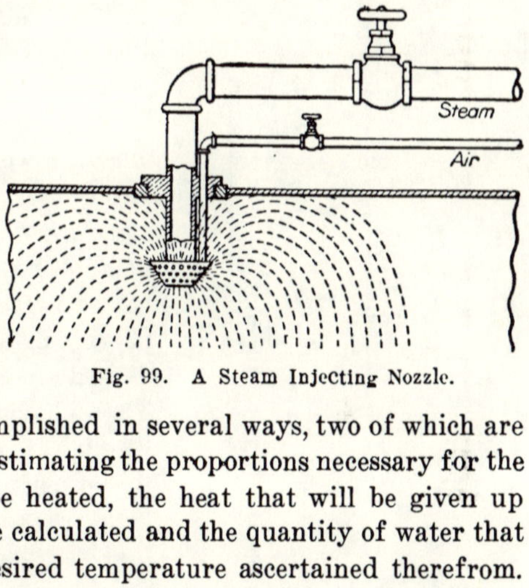

Fig. 99. A Steam Injecting Nozzle.

This may be accomplished in several ways, two of which are here described. When estimating the proportions necessary for the quantity of water to be heated, the heat that will be given up by 1 lb. of steam may be calculated and the quantity of water that will be raised to the desired temperature ascertained therefrom. The following table is not absolutely accurate, but may be used for the purpose of reaching approximately the sizes of boiler and pipes required, although the best plan to follow when cost of operation or other details requiring accuracy are demanded is to secure these from the basis of B.t.u. produced by the steam and absorbed by the water.

In the table the figures are based on a steam supply at 85 lb. pressure, and the quantities of water heated by 1 lb. of steam at that pressure are given in pounds:

Temp. of water to be heated F°	70	80	90	100	110	120	130	140	150	160	170	180
To boiling...	7	7½	8	8½	9½	10½	12	13½	16	19	23½	31
To 180° F...	8½	9½	10½	12	14	16	19	24	32	48	96	..

To use this table the initial temperature of the water to be heated must be found and subtracted from 212 deg. or 180 deg. Fahr. according as the temperature desired is one or the other. The difference in the temperatures is shown in approximate figures in the top line.

Under the steam pressure named, ⅛-in. pipe will inject into open bodies of water about 1 lb. of steam per minute.

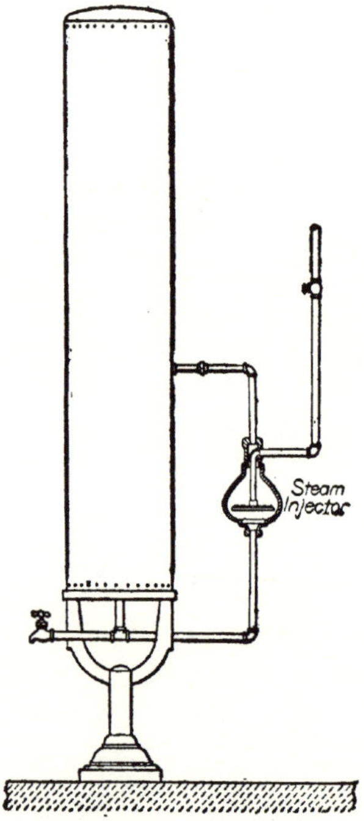

Fig. 100. A Special Steam and Water Mixing Attachment.

With steam from low-pressure mains (containing only about two-thirds as much heat in the liquids, a difference but partially offset by the greater latent heat) the difference in total heat and decreased velocity due to lower pressure will require much more time to perform an equal amount of work, say, in the neighborhood of 10 min. and less for pressures ranging from atmosphere, up, per pound of steam condensed at ordinary submergence. The delivery should be, of course, through a nozzle proportioned to the speed and volume or work in hand.

To ascertain correctly the amount of steam required to heat a given quantity of water per minute it need only be remembered that the heat unit is that amount of heat that will raise the temperature of water at 39 deg. Fahr. at the rate of 1 lb. 1 degree.

Finding Quantity of Steam Necessary to Heat Water.

The pressure of steam available should be ascertained and the total B.t.u. available from 1 lb. of steam will then be easily applied and the quantity of water it will heat found. For instance, steam at atmospheric pressure contains 1,140 B.t.u. per lb. Then if it was desired to heat the water from 40 deg. to 140 deg. Fahr. 100 B.t.u. would have to be transferred to each pound of water heated. Therefore one pound of steam would heat 11.4 lb. of water.

To find the quantity at higher temperatures and pressures of steam the following table of number of B.t.u. in one pound of steam at different temperatures may be used:

At atmospheric	pressure 1 lb. of steam contains approx.	1146	B.t.u.
At 5 lbs.	pressure 1 lb. of steam contains approx.	1151	B.t.u.
At 10 lbs.	pressure 1 lb. of steam contains approx.	1155	B.t.u.
At 15 lbs.	pressure 1 lb. of steam contains approx.	1158	B.t.u.
At 20 lbs.	pressure 1 lb. of steam contains approx.	1161	B.t.u.
At 25 lbs.	pressure 1 lb. of steam contains approx.	1163	B.t.u.
At 30 lbs.	pressure 1 lb. of steam contains approx.	1165	B.t.u.
At 35 lbs.	pressure 1 lb. of steam contains approx.	1167	B.t.u.
At 40 lbs.	pressure 1 lb. of steam contains approx.	1169	B.t.u.

When capacities of different apparatus are quoted in calories the corresponding capacity in B.t.u. may be determined as follows:—A calorie is roughly four times as large as a British thermal unit. For ordinary calculations this is the figure used; where more accuracy is desired the figure is 3.96. A British thermal unit represents the amount of heat involved in heating or cooling 1 lb. of water through 1 degree F. A calorie, or, as it is sometimes called, a greater calorie, represents the amount of heat involved in warming or cooling 1 kg. of water through 1 degree C. As 1 kg. weighs 2.2 lb. and 1 degree C. is 9/5 as large as 1 degree F., the calorie is $2.2 \times 9/5 = 3.96$ times as large as the British thermal unit.

When steam is introduced into a body of water in either a closed or open tank the condensation is accompanied by considerable noise due to the collapse of steam bubbles, the creation of vacuum at the point of collapse and the rushing in of water to that point to overcome the vacuum. To lessen this annoying accompaniment to the heating process the steam jet may be broken up and appliances are on the market which have been designed with the especial object of injecting steam so that the collapse of steam bubbles will be effected in such a manner, by reason of splitting up the jet into many minute jets, that much of the noise will be obviated. The manner in which this is often done in laundries, woolen factories and other institutions where open tanks are used for the purpose of boiling soap and for other purposes, and which are generally heated in this fashion, is to drill a large number of holes in the submerged copper or brass pipe, the end having been closed by plugging it. These lessen the noise greatly and it can be still further lessened by making a cage of perforated copper with fairly small perforations and fitting this around the part of the pipe which has been drilled for the steam emission. This cage simply acts to break up the force of the steam jets and to co-mingle them with the water more readily.

In a closed tank or boiler such as is used for domestic purposes, a type of steam jet is procurable which is designed to heat water noiselessly. This jet is in the form of an inverted cone which is attached to the end of the steam inlet pipe inside the tank and submerged in the water, as shown in Fig. 99. The steam is made to discharge outwards and upwards, thus causing a circulating motion in the contents of the tank and to avoid the noise resulting from the collapse of steam bubbles, a small air pipe carries air to the point where the steam escapes into the water. This air mixes with the steam and prevents the sudden collapse of the bubbles with the attendant noise. The steam and air supply are regulated to the proper proportions by valves. This heater may be used in a tank either under pressure or open, but when used as a closed system a pressure of air and steam in proportion to the pressure of water must be carried and check valves fitted on the inlet pipes to prevent water finding its way back into the steam boiler. Another apparatus, which is illustrated in Fig. 100, is fitted to the hot water boiler in exactly the same manner in which a gas heater would be fitted. From the illustration it will be seen that this takes the form of a closed vessel through which the water from the boiler may circulate and in which a steam jet with an outlet of special design is placed. As the water is heated it rises in the vessel and passes into the boiler, being replaced by cold water which enters through the lower connection. This process is assisted by the design of the jet, which breaks up the steam into minute particles and passes it into the vessel or co-mingler as it is termed in an upward direction, thus tending to hasten the speed of the circulation to the boiler. In this type of heater the tank may be under pressure if the pressure does not exceed that carried in the steam boiler and if check valves are placed on the steam supply line to prevent water from the tank finding its way into the steam boiler should the pressure fall.

Steam as an Auxiliary Heater.

A tank may be connected with both steam coils and a water heater of the usual type as shown in Fig. 101. This is commonly done where steam is available from the heating system in the winter time while the water may be heated from the tank heater

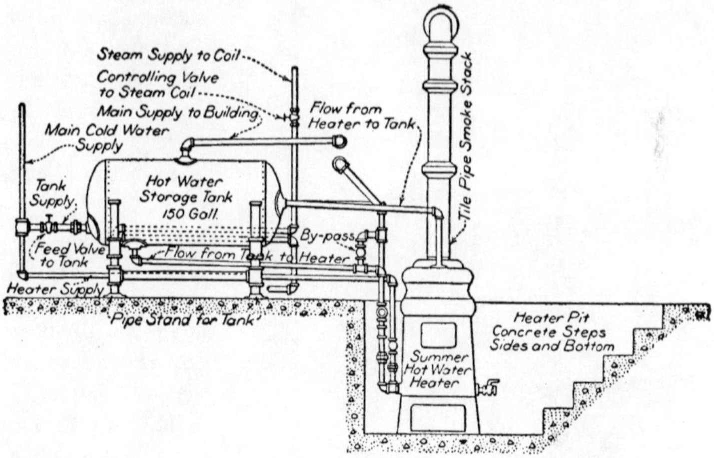

Fig. 101. Method of Connecting Steam Coil and Heater to Hot Water Tank.

at such times as the heating system is not in use. In the illustration the heater is placed in a pit but this need only be done when constructional obstacles prevent the raising of the boiler on a stand to such height as will allow of a return to the heater by gravity of the return water. Valves are also shown on the supply to both heater and tank, these being intended to enable the cold supply to the heater to be closed at such times as the water is being heated by steam, and so when the by-pass valve is opened and that on the return connection lower down is closed, cause the circulation to the building to take a shorter path on its return to the storage tank. When the water is being heated by the tank heater the reverse is the case, as then the valve on the by-pass is closed and that on the return connection opened while the cold supply valve at the tank is closed and the feed water caused to enter the tank heater by the connection to the return pipe as it enters the boiler. A boiler stand built of pipe is shown in the illustration as well.

Auxiliary Hot Water Heaters.

The economy attendant on the use of auxiliary hot water heaters in the form of coils or specially designed vessels which may be placed in the fire box of a steam, hot water, or warm air furnac without interference with its efficiency for its main purpose is apparent. Where such provision is made for heating the water for domestic purposes during the season that the heater is in daily use, and an automatic gas water heater is used during the remainder of the year, a very high degree of efficiency in the service is obtained. Most heaters are now provided with a passage way for the circulating pipes that connect such water heaters. These generally are placed at the side of the fire box door and are covered with a plate which is easily removed should the connection be desired. Fig. 102 shows such a heater installed in a Kelsey warm air furnace. In this case the flow pipe is carried down through the monitor top of the warm air chamber, then through the dome of the fire box and is connected into the top of a cast iron water heater of special design, which exposes considerable surface to the radiant heat of the fire. The

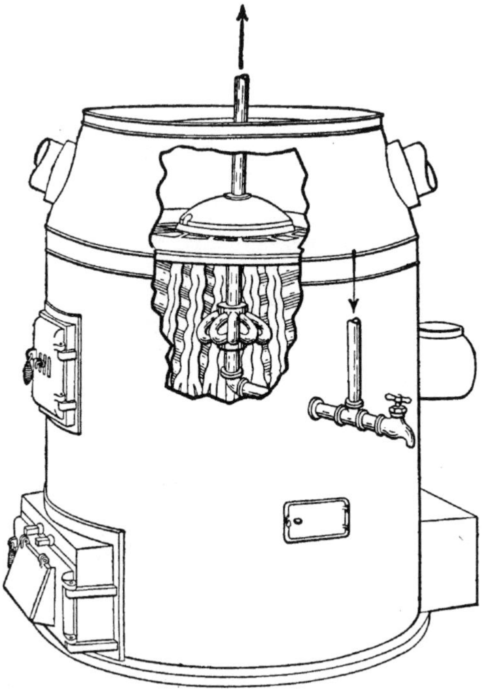

Fig. 102. Auxiliary Heater Fitted in Warm Air Furnace.

return connection is made by drilling a hole in the flange which covers the junction of the vertical warm air passages and passing the pipe through this. A gas tight joint is made by careful fitting and the provision of a locknut on the outside surface of the air passages where it passes through the fiues. The connection may also be made by taking the return back through the dome top, but in such a case there is no means of draining the heater when it is out of use for any reason.

Instead of the cast iron heater a coil of brass or iron pipe is frequently used, this being made to fit closely around the walls of the fire box and as a rule being placed at such a height that at least one pipe is in contact with the fire. Such a coil is shown in Fig. 103 fitted to the fire box of a round sectional boiler, the pipes passing through the plates at the side of the fire door by means of openings specially provided. Such coils are easily built from brass pipe. In constructing them annealed or semi-annealed pipe should be selected, as it may be bent more easily and a neater job will result. The length of the coil should be ascertained by bending a wire to the desired shape, taking care that the return bend will be as close to the fire door as possible so that the maximum benefit will be secured. The pipes should then be cut and threaded and screwed into the return bend, when the two can be bent together by passing them between two firmly

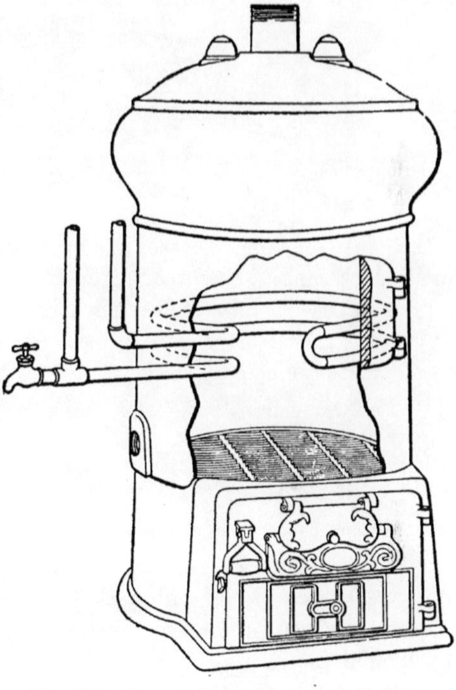

Fig. 103. Brass Water Heating Coil Fitted in House Heater.

fastened planks and bending a little at a time until the desired shape is secured. Care must be taken to secure a pitch from the lower connection clear around the coil until it passes through the walls of the heater again, else there will probably be complaints as to noise in operation through steam collecting in the coil. If the fire box is a large one some provision may be necessary to support the end of the coil at the return bend and the ingenuity of the fitter must be exercised to secure this. Another type of heater is shown in Fig. 104. This may be fitted with ease instead of the coil described and may be placed in the fire box of any type of heater where it is possible to secure the passage of the pipes. The method of making the connection to the boiler under various circumstances is described elsewhere. Such coils and heating apparatus are also frequently used for the supply of hot water to a radiator in some apartment too far from the heater to warm satisfactorily with hot air. In such cases the rating given them is about as follows:

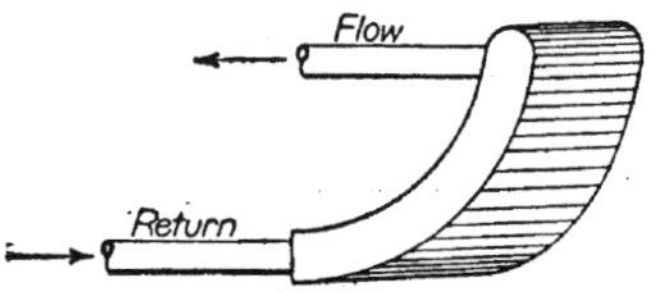

Fig. 104. A Heater Adaptable for Nearly all Furnaces.

Cast iron sections suspended from fire box crown	1 sq. ft. to 15 sq. ft. radiation
Cast iron sections in contact with fire	1 sq. ft. to 50 sq. ft. radiation
Pipe coil above fire	1 sq. ft. to 20 sq. ft. radiation
Pipe coil partly in contact with fire	1 sq. ft. to 35 sq. ft. radiation

As these ratings are based on a flow of water to the radiators at 170 deg. Fahr. and as that is about the highest temperature maintained in hot water supply tanks, it may be calculated that for domestic use:

1 sq. ft. of cast iron surface suspended over fire will heat 10 gal. per hour
1 sq. ft. of cast iron surface in contact with the fire will heat 25 gal. per hour
1 sq. ft. iron coil surface suspended above fire will heat 15 gal. per hour
1 sq. ft. iron coil surface partly in contact with fire will heat 25 gal. per hour
1 sq. ft. coil surface, brass pipe coil, partly in contact with fire will heat 30 gal. per hour

These figures are approximate only, having been compiled from observation of results obtained under ordinary working conditions rather than from the amount of heat absorbed and transferred by the heating surface, from contact with the fuel or by radiation therefrom.

If it is desired to calculate the heat that should be transferred to the water and the amount of water that should be heated to the desired temperature by a given amount of surface, it is safe to allow a heating value of 15,000 heat units per sq. ft. of coil, and for water heaters suspended above the fire a much lower rating, possibly 8,000 heat units is an ample allowance for the ordinary conditions found in heating boilers and furnaces.

Table of equivalent heating surface in pipe coils:

1 ft. of ¾ in. pipe=0.275 sq. ft. surface	45 in. of ¾ in. pipe=1 sq. ft. surface
1 ft. of 1 in. pipe=0.346 sq. ft. surface	35 in. of 1 in. pipe=1 sq. ft. surface
1 ft. of 1¼ in. pipe=0.434 sq. ft. surface	28 in. of 1¼ in. pipe=1 sq. ft. surface
1 ft. of 1½ in. pipe=0.494 sq. ft. surface	24 in. of 1½ in. pipe=1 sq. ft. surface
1 ft. of 2 in. pipe=0.622 sq. ft. surface	20 in. of 2 in. pipe=1 sq. ft. surface

Besides the coils of copper or brass pipe and the cast iron heaters which are intended to be hung over the fire in warm

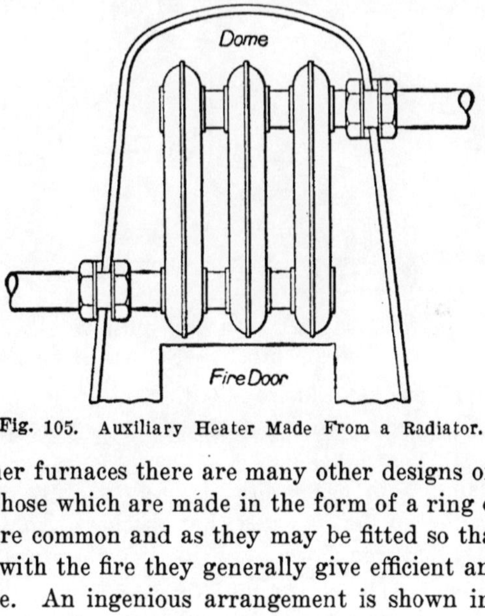

Fig. 105. Auxiliary Heater Made From a Radiator.

air and other furnaces there are many other designs of auxiliary heaters. Those which are made in the form of a ring or the part of a ring are common and as they may be fitted so that they are in contact with the fire they generally give efficient and satisfactory service. An ingenious arrangement is shown in Fig. 105, where a heater of considerable power and heating value is constructed from the sections of an ordinary hot water radiator. The heater is intended to be placed in the dome of a warm air furnace and is made of radiator sections to the number and of such height as the size of the furnace and the nature of the work performed demand. If a heater is built up of 3 columns 18 in. high, there will be a heating surface of about 7 sq. ft. available. As this is not in direct contact with the fire the full transmission rate of cast iron heating surface cannot be applied to it, but if a rate of 10,000 heat units per sq. ft. is allowed this will mean a transmission to the water of 70,000 heat units per hour when a strong fire is maintained. This is equal to raising 700 lb. or 84 gal. of water from 50 deg. to 150 deg. per hour. If the fire in the furnace be run at a low rate and only maintained at a dull red heat the transmission would possibly not be over 5,000 B.t.u. per sq. ft. when of course the water heated would be 42 gal. The allowance of 10 gal. per sq. ft. may be taken as a good average for this also. The laundry stove, which combines

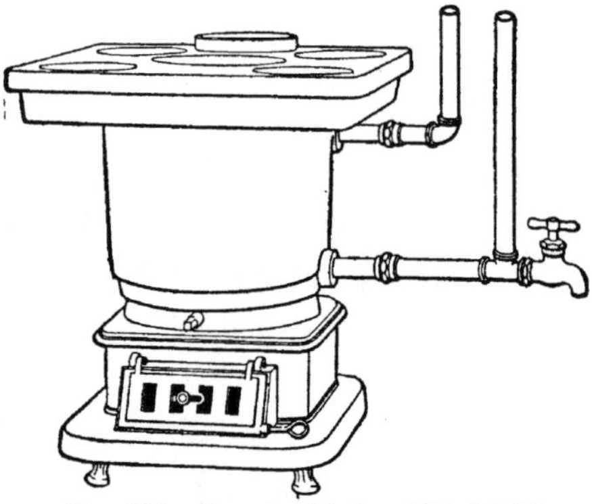

Fig. 106. Laundry Heater Fitted With Water Section.

the purpose of heating sad irons with that of heating water, is shown at Fig. 106. This is a common example of auxiliary heater.

In arranging to heat water by an auxiliary heater in a furnace it must not be expected that the heat that is being obtained is heat that would otherwise go altogether to waste. Each pound of fuel consumed has only so many heat units to give up, and if it is transferred to the water for domestic or heating purposes it stands to reason that it must rob some other part of the heat that would have gone there. If, however, the heater is of ample size for the work it has to do, there is economy in the use of such a system of water supply as the heat losses in the flues remains the same.

CHAPTER XII.

Utilizing Excess Heat in Heating Rooms and Domestic Appliances.

In many cases the kitchen range or laundry heater which also supplies heat for the hot-water supply system is in use for long periods when no hot water is required or it may by reason

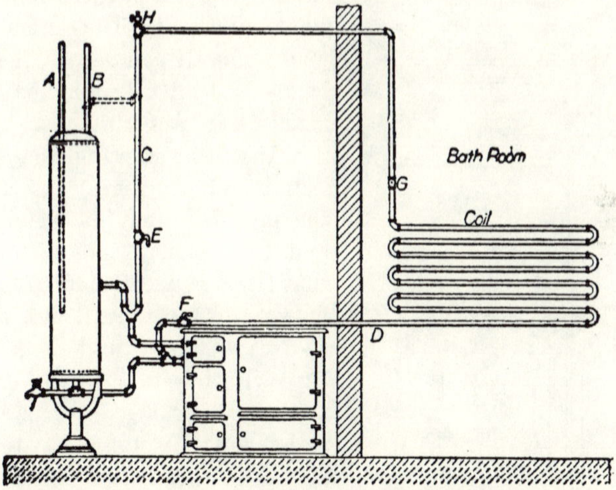

Fig. 107. Coil on Same Floor as Boiler Heated From Water Back.

of the large demand on its capacity be of larger size than the supply of water required would call for. In this case there is a strong possibility of the water becoming overheated and causing annoying rumbling sounds each time the faucets are opened, owing to the water at high temperature flashing into steam when the pressure is temporarily lowered. One method of utilizing this excess heat is to connect the system with a radiator in some one of the rooms and it can thus be taken advantage of in the winter season while it is not likely that the range will be used to such extent in summer as to make the water excessively hot for long continued periods. The size of radiator that may be added without seriously affecting the supply of hot water may be estimated by allowing 2 sq. ft. of surface for each gallon of water that the water front will heat.

The methods of connecting the radiator or coil are similar to that which would be followed on an ordinary hot-water heating system. Where it is desired to heat a room on the same level as the range the connection is somewhat different. The manner in which a wall coil set on the same level as the boiler may be heated is shown in Fig. 107.

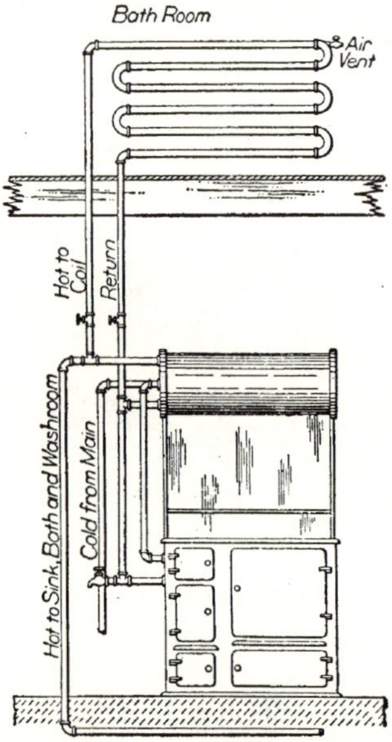

Fig. 108. Coil on Floor Above Boiler Heated from Water Back.

It will be seen that the connection is made by means of a Y on the pipe between the water front and the boiler. This is done so that it may be switched off entirely in warm weather and also so that the water at its highest temperature will flow to the coil. It however can be connected with equally satisfactory results at the top connection of the boiler, as shown by the dotted line. At the highest point an air cock is fitted to relieve the loop of air collecting there. If the system is supplied from an overhead tank this air cock can be replaced by a pipe which is taken to a point higher than the tank and then returned to a point over it so that any water expanded back through it will be delivered into the tank. No air cock is required on the coil as the cock or air pipe on the high point relieves the coil of air also. The return pipe is connected to the return connection of the water front as shown and a stop cock on this pipe also effectively prevents water backing up into the coil at such times as its use is not desirable.

Connecting Coils on Floor Above Range.

The circulation of water to a coil or radiator on the floor above the boiler is a simple matter. Two methods of connecting them are here shown. In Fig. 108 the circulation between the back and boiler is fully maintained and is direct, while the circulation between the boiler and heating coil or radiator is also secured, the descending or cooler column—that is, the return water from the heating coil—being arranged to flow in the same direction as the same column from the boiler to the back. In Fig. 109 the flow pipe, or ascending column, is taken directly from the water back to the heating coil, on the horizontal part of which pipe is a circulating pipe, to the boiler. When starting the fire, the circulation through the pipe to the heating coil will be more rapid than through that to the boiler. The water in the boiler will become heated by the return water from the heating coil, as well as the partial circulation through the pipe. As the temperature of the water in the boiler increases, the circulation through the heating coil will decrease, and if the temperature of the water in the boiler should become equal to the temperature of the water in the coil, which is not improbable, the circulation in the heating coil will cease. If a valve was placed on the circulating pipe and closed when the coil was used, the circulation would be continuous and the water in the boiler would be slowly heated by the return water from the coil. When hot water is drawn off for domestic purposes, if the pipes are arranged as shown in Fig. 109, the temperature of the heating coil will be immediately reduced, as the hot water in the coil is as liable to be taken off as that in the boiler, whereas in Fig. 108

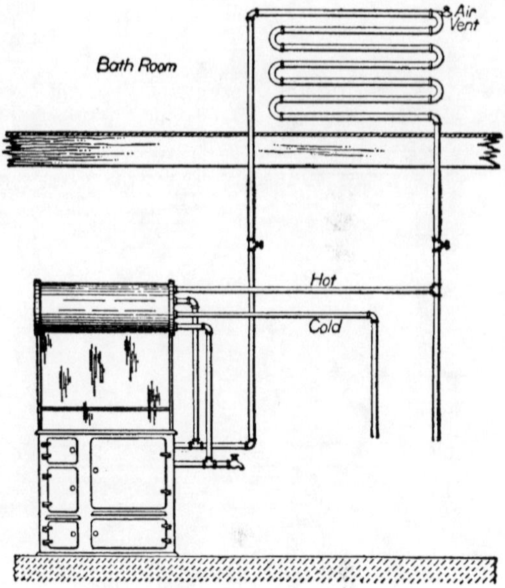

Fig. 109. Connections Direct From Water Back to Coil.

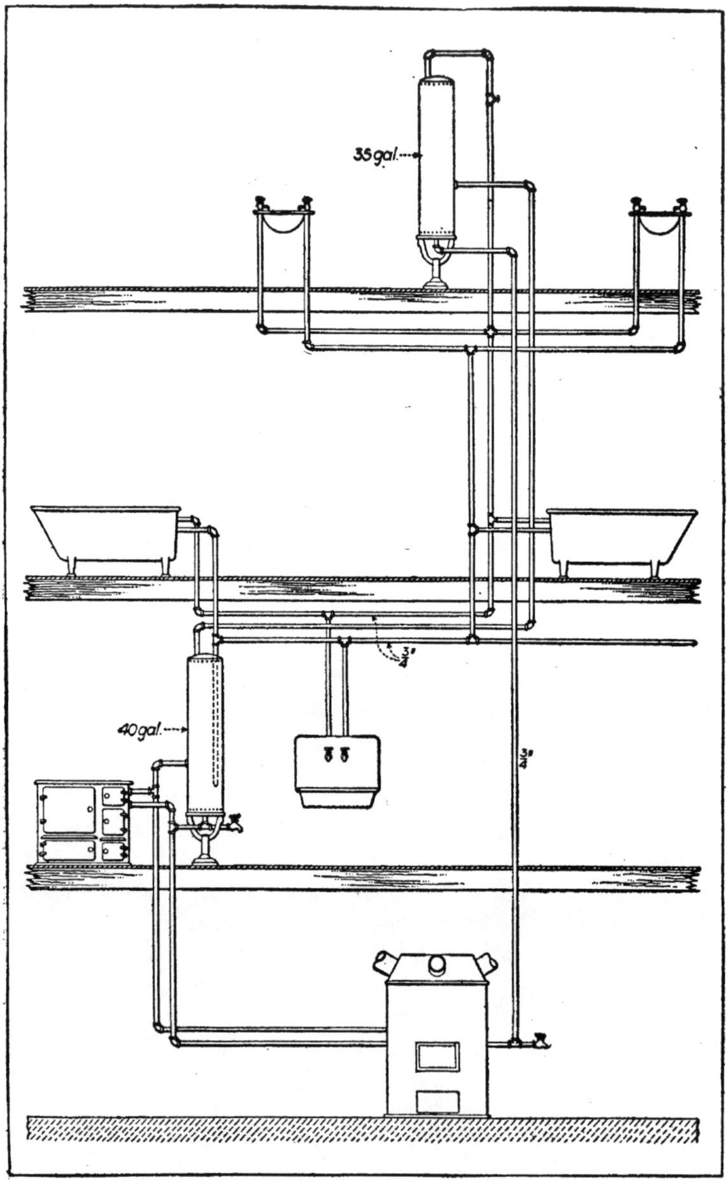

Fig. 110. Hot Water Boiler Placed in Attic as Storage and Radiator.

the liability to withdraw the hot water from the coil does not exist, and the incoming cold water only retards the supply of heat to the coil. The difference between the two plans may be thus summarized: In Fig. 108 the heating of the domestic water supply in the boiler is not interfered with, and the circulation through the heating coil is not liable to be reduced or to cease; whereas in Fig. 109 the heating of the water in the boiler is liable to stop the circulation in the heating coil, and the withdrawal of hot water for domestic purposes and the incoming cold water will tend to cause irregular circulation between back and coil, and back and boiler. For the radiating surface a return bend coil may be used of $1\frac{1}{4}$-inch pipe. The length of coil may be about 3 feet 6 inches, eight to ten pipes in height, but the size of the room and the capacity of the water back must determine this. The flow and return pipe may be of $\frac{3}{4}$-inch pipe, if the coil is not unusually far from the boiler. The use of plain or galvanized iron pipe is a matter which may be determined from experience when considering the requirements of the case. So far as heating the air is concerned, plain pipe is more desirable than galvanized pipe, but the latter pipe is presumed to be less liable to produce rust in the water for domestic purposes. There is no objection to the use of a radiator if pre-

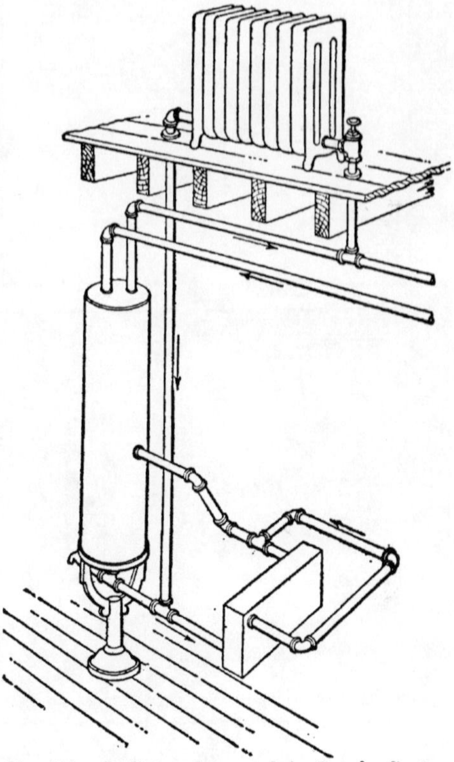

Fig. 111. Radiator Connected to Supply System and Extended Heating Surface at Water Back.

ferred. When these are used on the same floor as the boiler, the wall pattern should be selected so that the return connection may be kept above the level of the water front. Better service will result when this is done.

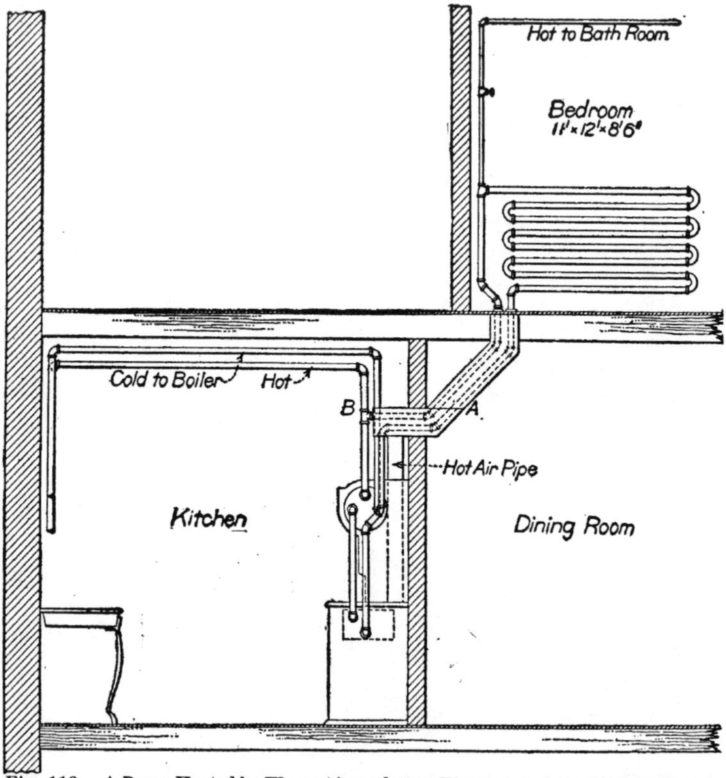

Fig. 112. A Room Heated by Warm Air and Hot Water from the Kitchen Range.

The equivalent in pipe coils of 1 sq. ft. of radiation may be found from the following table:

LENGTH OF IRON PIPE REQUIRED TO EQUAL 1 SQ. FT. OF RADIATING SURFACE.

¾ in.	About 3 ft. 8 in.
1	About 2 ft. 11 in.
1¼	About 2 ft. 4 in.
1½	About 2 ft. 0 in.
2	About 1 ft. 8 in.

Warming a Room by Installing an Extra Boiler.

Another plan often used is to install an extra boiler instead of a radiator or coil and to place this in some room that requires heat so that the radiation from its surface will effect this purpose while the amount of water held in store for immediate use is much increased. An example of this method is shown in Fig. 110, where an attic bedroom has been warmed by the circulation of water to an extra boiler placed therein. In this case the water is heated by a coil in the warm air furnace as well as by the water front in the kitchen range and the circulation of both coil and water back is primarily to the first boiler. The flow to the upper boiler is taken from the top of the lower one and the return enters the heater section or coil in the hot air furnace so as to maintain a circulation throughout. It could, however, have been connected to the return pipe below the upper boiler with satisfactory results.

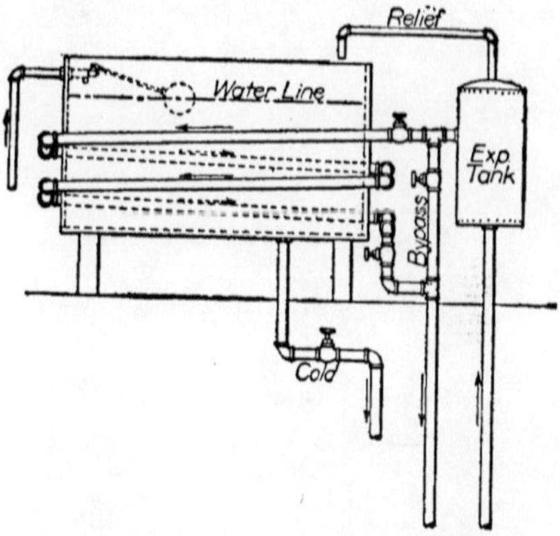

Fig. 113. Cold Water Storage Tank Provided With Warming Coil Heated by Water Back.

Connecting a Radiator to Domestic Supply Lines.

A little variation in the method of connecting a radiator on the floor above the boiler is shown in Fig. 111. The flow pipe here is taken from the supply pipe to the fixtures and returns through a separate pipe to the lower connection of the range boiler. An air valve is necessary on the radiator to relieve it of the air collecting there. The manner of extending the surface of a water front to provide extra heating surface for a large radiator is also shown. This has been described elsewhere in this book. The valve may be fitted on the return connection if preferred, but there will probably be some local circulation in the radiator if this is done.

Kitchen Stove Warming Room by Hot Air and Hot Water.

An ingenious arrangement for the utilization of excess heat from a kitchen range is shown in Fig. 112. This range has a special construction which permits part of the heat of combustion, after doing duty in warming the ovens and water front to be utilized in heating air which is delivered to a near by room in the same manner that warm air is delivered from a regular heating furnace. In this case, however, advantage was taken of the warm air duct to run the hot water pipes to a coil in the room to

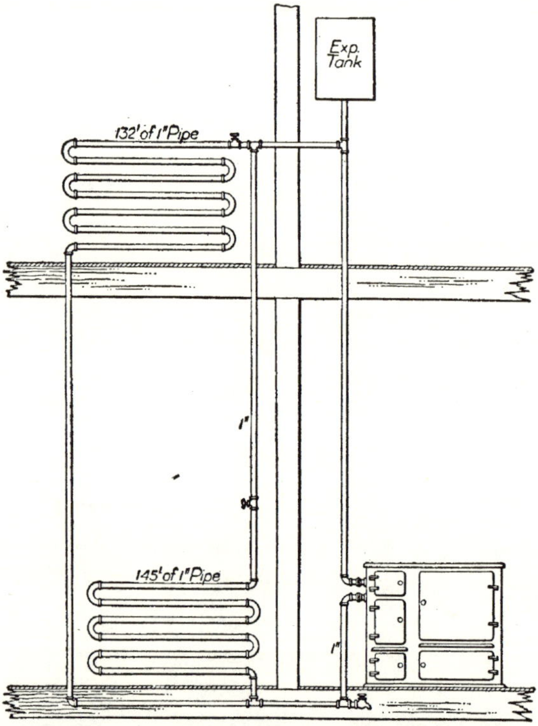

Fig. 114. Two Coils on Different Floors Heated by Water Back.

which the warm air was conducted. The illustration shows how the pipes were connected to a horizontal boiler placed above the range and how the pipes passed up through the floor register to the coil. As a means of relieving the system of air a pipe connecting with the bathroom fixtures was taken from the highest point.

There is still another useful purpose to which the warm water not required for domestic purposes can be put. In many instances the cold water storage tank is placed in an unheated attic and in winter there is considerable danger of the pipes and tank freezing. It is an easy matter to construct a coil around the tank in the manner shown in Fig. 113, and to connect this so that there will be a circulation of water to it at all times which will guard against freezing even when the temperature is at an extremely low point. If the expansion tank is placed as shown, there will be sufficient water always therein to give a little head of water above the coil, as the water when hot will stand up in the tank a little higher than the level of the water in the cold water storage tank owing to the difference in density. If connected as shown the supply to the coil may be cut out in mild weather without affecting the circulation in the rest of the system.

Warming Several Rooms From the Kitchen Stove.

In such cases as those in which the water for domestic purposes is entirely heated by steam or by a special heater the excess heat of a kitchen range not required for cooking is often utilized for heating rooms. This does not require the use of a boiler, as the water front then simply takes the place of the ordinary house heating boiler, although of course few water fronts have capacity enough to heat other than a small room. Fig. 114 shows the water front in a kitchen range heating two coils, one of which is on the second floor. The circulation is such that either or both coils can be in operation without affecting it and the expansion tank at the highest point of the loop takes care of the expansion in the system and also stores water enough to allow of a certain amount of evaporation without constant attention in refilling.

Fig. 115 shows two radiators on the second floor connected in such a manner that they may be heated either by the kitchen range or by the coil in a warm air furnace or by both. The connections to the radiator are made in each case so that the supply comes from overhead, obviating the use of an air valve and allowing the water to be circulated through either or both radiators as desired, or merely around the circulating loop without warming the radiators. The same method can be used in supplying radiators on a system from which the domestic supply is taken, but in that case an expansion tank would not be required as the relief pipe would be turned over the top of the storage tank in the attic and a boiler would be required to maintain a constant

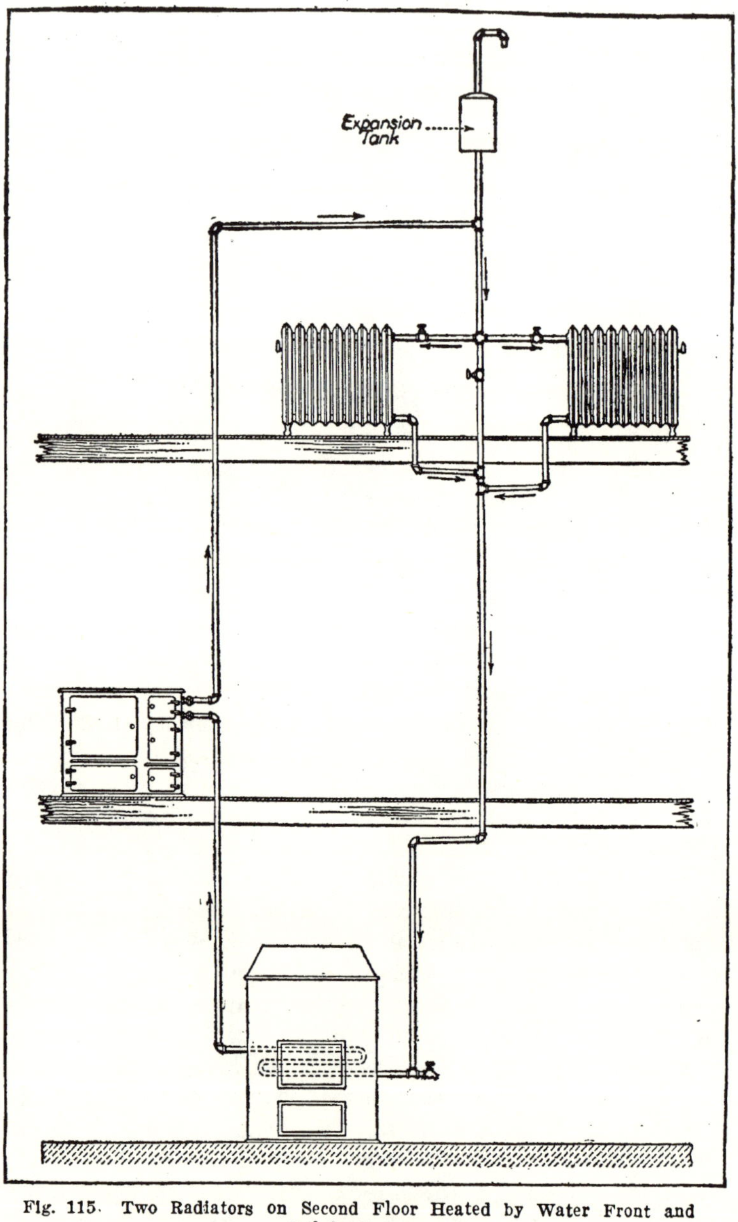

Fig. 115. Two Radiators on Second Floor Heated by Water Front and Coil in Furnace.

supply of hot water for use at the fixtures. The connections would then be made as shown in Fig. 116, and the cold supply would of course be constant instead of being merely filled as the evaporation lowered the level in the system.

A Plate Warming Closet Heated by Hot Water.

One of the greatest conveniences in a residence is a warming closet or table in the butler's pantry or kitchen. This enables food to be kept warm or plates and other dishes kept warm for serving the food, and such a closet may easily be warmed by the water from the kitchen range boiler.

Fig. 117 shows how the water may be circulated to various fixtures so that hot water is available instantly at the opening of a faucet, while the warming closet is heated by the return pipe. This can be by-passed as shown by the dotted lines if desired, so that the closet may be left unwarmed at will. The fixtures on the upper floors should be connected to the supply pipe so that the act of opening a faucet would remove the air collecting at the highest point, or if a tank supply is used the pipe should be vented from the highest point. If all the connections are made in the manner shown there will be little chance of water being drawn through the return and so reversing the flow for the time being. If, however, a connection must be made lower than the coil and there is a possibility of the water being drawn back from the return connection instead of the flow, a light check valve may be inserted. This can be inserted in such a fashion, by pitching a section of the return pipe, that the swinging check will hang almost vertical and will therefore offer very little obstruction to circulation, while it will close immediately the flow is reversed. The manner of hanging the check valve is shown in Fig. 118.

Heated Towel Rails.

A fixture of considerable utility in the bath room, and one, moreover, that can be made to have an exceedingly attractive appearance, is the heated towel rail; yet it is one that has been almost entirely overlooked by plumbers. The ordinary bent tube as a convenience for hanging towels upon is good enough but the comfort and convenience of having always warm and dry towels in the bath room are so obvious that when these fixtures

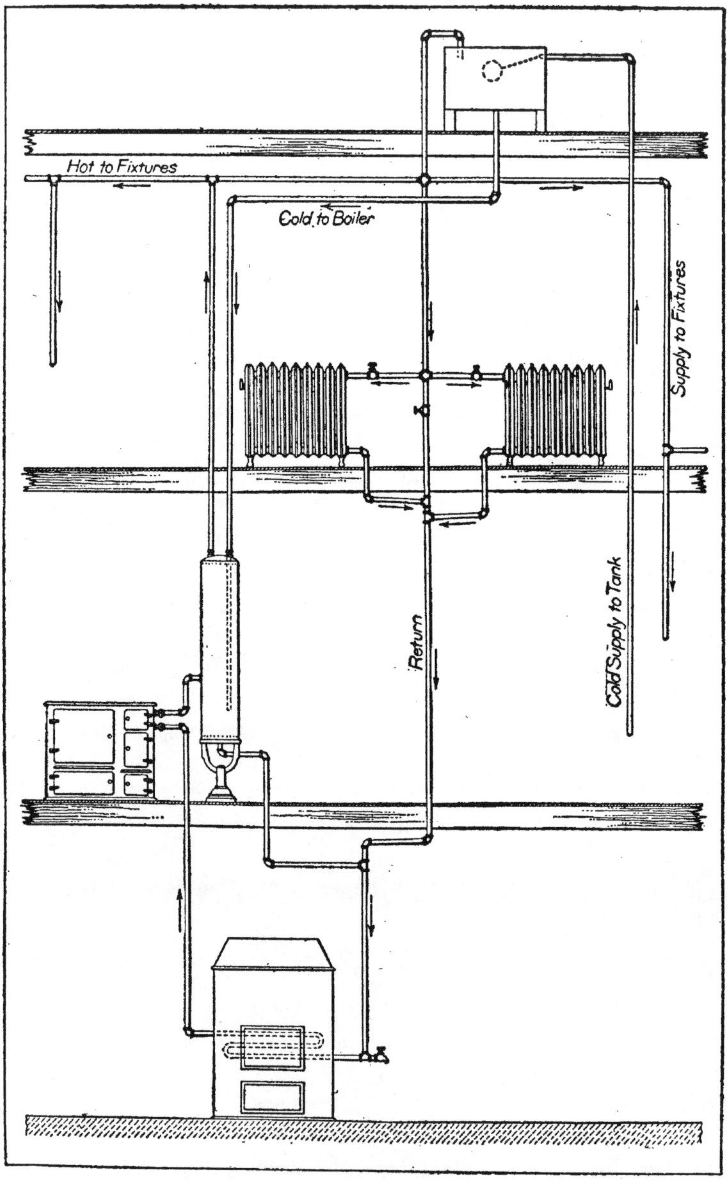

Fig. 116. Radiators Fitted in Connection with Regular Domestic Supply System.

are shown to a prospective customer a sale will result almost every time.

Towel rails heated by a circulation of hot water from the ordinary domestic supply are commonly used in England. In fact, no first-class bath room or toilet room would be considered complete without one, and there will probably be some considerable use made of this fixture in America now that American manufacturers are including them in their catalogues. The pat-

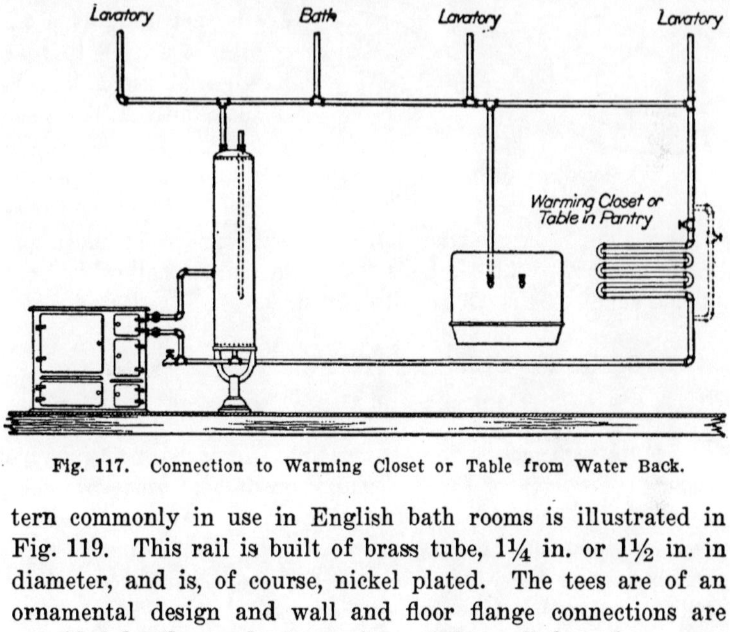

Fig. 117. Connection to Warming Closet or Table from Water Back.

tern commonly in use in English bath rooms is illustrated in Fig. 119. This rail is built of brass tube, 1¼ in. or 1½ in. in diameter, and is, of course, nickel plated. The tees are of an ornamental design and wall and floor flange connections are provided for flow and return pipes. Where all four flanges are fixed to the wall the controlling valve is generally placed between the wall flange and the rail and, of course, the flow may enter the top rail instead of the bottom if preferred. This is sometimes desirable and, of course, necessary if the supply comes from overhead.

A larger rail, designed to stand clear out in the room, is also used. This is called a double rail and the name is descriptive of the fixture, as it is simply two rails cross-connected. The connections to these are, of course, always in the floor. The method of connecting a towel rail is identical with that of connecting a radiator, and there is no trouble in heating them at all if the position is such that a circulation is possible.

The size of these towel rails varies according to the designs of the different makers, but 30 in. wide is found generally suitable for a three-tube rail. A towel rail to give satisfactory results, even if it has not so good an appearance, can be constructed from the ordinary stock pipe fittings. One-inch brass pipe, tees, right and left ells, petcock and a few nipples are all that are required, and if carefully built quite a good-looking fixture can be produced. Solid nipples are used in the flanges that are not to be used for flow or return connections, and a petcock or air valve fitted at the top of the rail will allow for occasional relief of air. This rail is illustrated in Fig. 120.

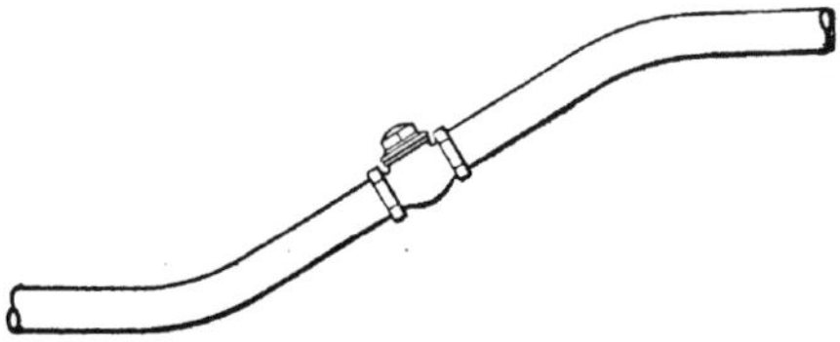

Fig. 118. Method of Fitting Check Valve to Allow it to Swing Freely.

Utilization of Waste Heat in Heating Water for Domestic and Other Purposes.

Many schemes have been brought forward in the way of securing economy in fuel by utilization of heat otherwise going to waste for the purpose of heating water. Among those may be mentioned that of using the waste heat of a gas or oil engine by circulating the water from the cooling jacket through a coil which is placed in the water storage tank. This is applicable in any factory or other place such as a creamery, where a gas engine is run for any length of time. Even if the heat generated by the engine is not sufficient to warm all of the water required to the desired temperature, it will at least be enough to sensibly raise the temperature and thus save fuel by causing the initial temperature from which the heater has to raise the water to be considerably higher. The condensation from steam traps, hot closets and other manufacturing appliances where the water of condensation is not returned to the boilers may be used advantageously in a similar manner. These are not applications of heating practice that lend themselves readily to illustration, as each case must be treated according to local conditions, but one application of the idea is seen in Fig. 121. This installation is in a steam laundry and there are so many institutions where large quantities of hot water are running to waste that it would seem that whenever large quantities of hot water are used and discharged into the sewers this scheme might be profitably adopted. A very considerable saving in fuel is effected and the necessary coil and tanks are simple and comparatively inexpensive to install. It also has the advantage of reducing the temperature of the wastes entering the sewers, and this is eminently desirable.

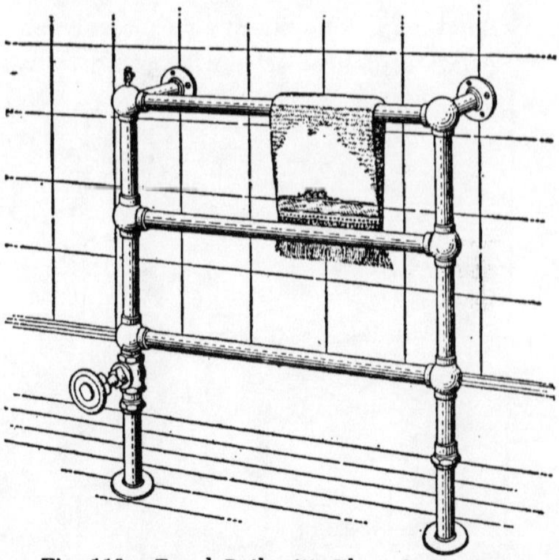

Fig. 119. Towel Rail with Floor Connections.

In Fig. 121 is shown the supply to a hot-water tank in a steam laundry passing through a coil made of about 800 ft. of 2-in. galvanized iron pipe. This coil is made in two sections, one being placed in each compartment of the concrete tank as shown in plan in Fig. 122. The object of making the tank double is to insure the waste water passing through its entire length, connection between the two being made only at the ends furthest from inlet and outlet pipes by means of 4-in. thimbles in the partition wall. Into this tank is discharged all the hot waste water from the washing machines, and as the temperature of this will average around 180 deg. and the heating surface of the coil is approximately 500 ft., it will be understood that the temperature of the supply is raised to a very considerable extent in passing through it.

The pipe from coil to boiler is covered, as is also the boiler, with asbestos covering, and the steam supply to coil heater in the boiler is thermostatically controlled. Thus heat is conserved at both ends and the amount of steam required to maintain a supply of hot water to the laundry has been reduced to an extent that has made the installation of the coil and tanks worth while.

Water Heating by Garbage Burning.

Another development in the economical use of fuel and the conservation of heat otherwise going to waste is seen in the combination garbage incinerator and water heater. These heaters are productive of considerable economy in many cases as the garbage and wastes of the building are often of sufficient quantity to provide a large proportion of the heat necessary for the hot water supply. The construction of one type of these garbage burning water heaters is shown in Fig. 123. It will be seen that the garbage does not come in actual contact with the fuel on the grate of the fire box, but is contained in a receptacle immediately above it. This has a grate formed of pipe connected to the double shell of the heater. Thus besides the heating surface afforded by the walls and dome, the pipe which forms the receptacle holding the garbage also acts as heating surface. The garbage dries over the fire and is gradually consumed, giving up its heat to the water in the process and where a good draft is available the process of combustion is carried out very completely. Such a heater is proving of value as a ready means of disposing of the waste paper, straw and other litter of shipping rooms in business houses, and the garbage and household wastes of apartment buildings and hotels. The connections to the tank are made in the same manner as with an ordinary tank heater, and the drafts may be controlled by a thermostat or by hand as desired.

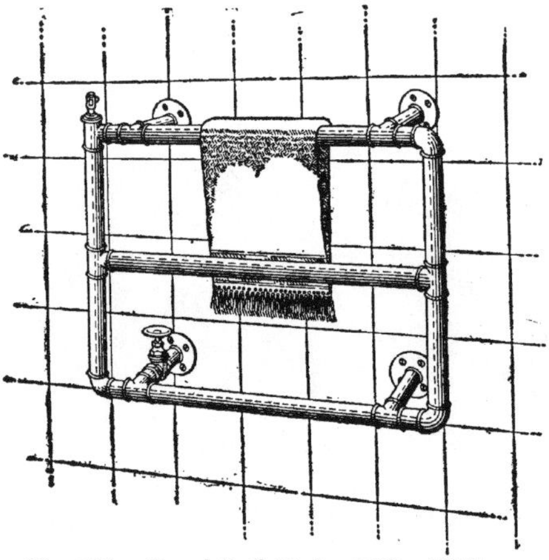

Fig. 120. Towel Rail Made of Pipe Fittings.

Advantage is generally taken in bakeries of the heat of the sand which covers the ovens to retain their heat in warming the water used in the work of the employees. In Fig. 124 is shown a common method of heating water. In this case the boiler is simply buried in the sand and the supply connection made to the side outlet. The cold water supply enters the boiler through

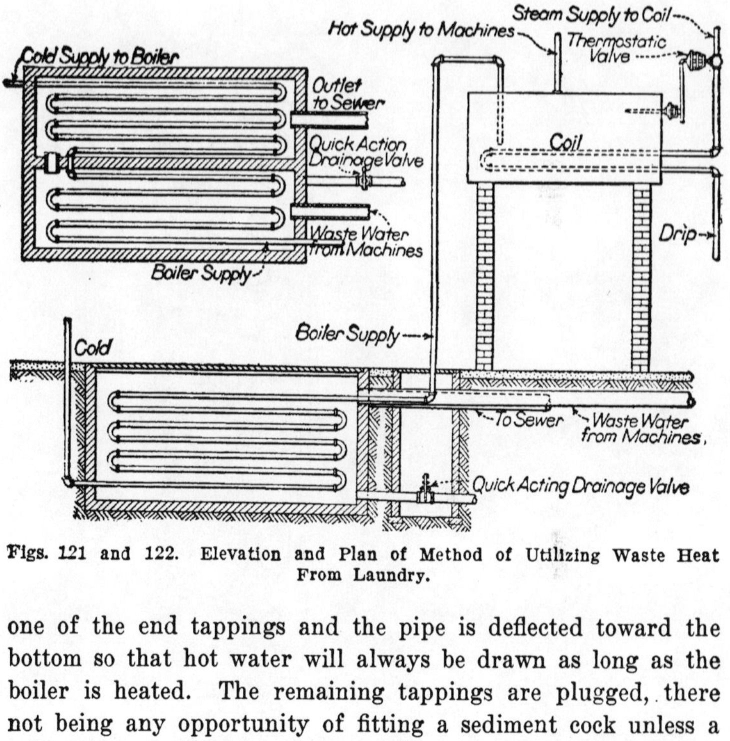

Figs. 121 and 122. Elevation and Plan of Method of Utilizing Waste Heat From Laundry.

one of the end tappings and the pipe is deflected toward the bottom so that hot water will always be drawn as long as the boiler is heated. The remaining tappings are plugged, there not being any opportunity of fitting a sediment cock unless a boiler with a special tapping on the side is secured, or a dip pipe fitted as shown, when a pipe may be brought out through the brickwork and a sediment cock fitted to the end of it.

Heat is transmitted to the boiler from the sand and an ample supply of hot water is generally available. This system, however, has the disadvantage of being difficult to get at for minor repairs and in localities where much sediment is deposited or where the water tends to corrode the boiler rapidly the alternative system shown in Fig. 125 is preferable. This has a copper or brass coil running along one side of the oven, the pipes being brought through and connected to the boiler in the usual manner. The boiler may be placed higher than usual to admit of the connection being made to the side tapping, as shown, or it may be placed at the usual height and a connection made to one of the top tappings. In either way it is an eco-

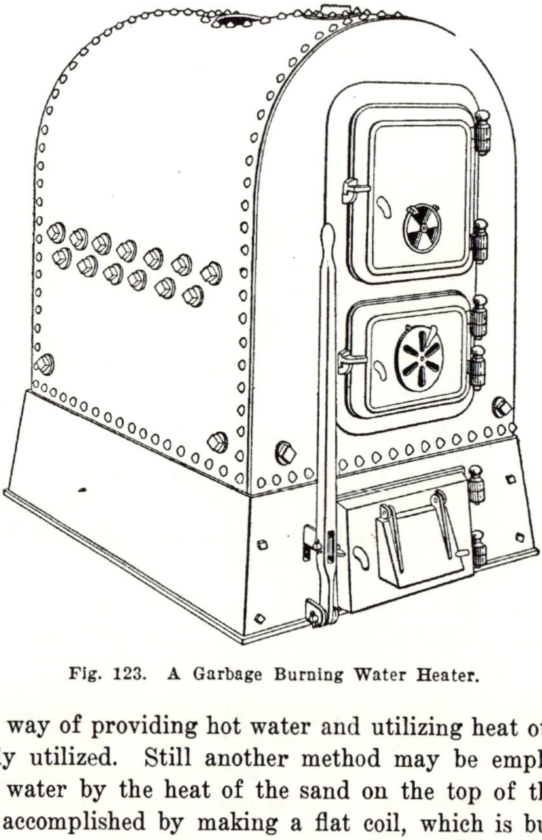

Fig. 123. A Garbage Burning Water Heater.

nomical way of providing hot water and utilizing heat otherwise not fully utilized. Still another method may be employed in heating water by the heat of the sand on the top of the oven. This is accomplished by making a flat coil, which is buried in the sand and connected to the boiler in the usual manner. In most cases it will be necessary to use a horizontal boiler and to place this at a height which will admit of making the connections to it so that a pitch will be secured from the coil to the side or end connection of the boiler.

It is well to make the coil rather long or to use pipe of

large diameter, $1\frac{1}{4}$ in. for preference. This is necessary owing to the lower heat of the sand than that available when the coil is placed inside the oven, and to the fact that the absorption of heat from the sand tends to keep the layers next to the pipe at a somewhat lower temperature than the rest. The draw off cock in this case is fitted to the coil and brought to a point outside of the brickwork.

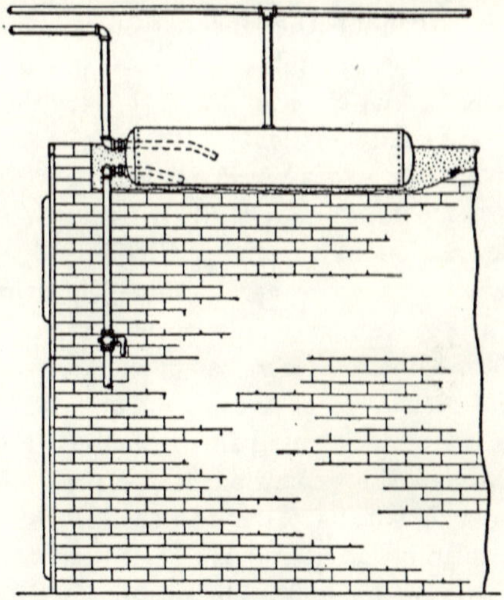

Fig. 124. Method of Heating Boiler by Placing it in Sand Over Oven.

Care should be taken to lay the coil so that there will be no opportunity for air to collect and so that it pitches upward continuously to the boiler connection and down from the return connection to the draw off cock. Brass pipe is best suited for such a coil and it may be made either with iron headers or return bends to avoid spreading the fittings in making the joints. Where it is desired to supply one large boiler with hot water from coils in several ovens it may be well to connect them into a header and from there make the connection to the boiler. The returns in such a case would also be connected through a header to the lower tapping of the boiler. Care must be taken in doing so to make the connections so that the flow from each coil would be equalized.

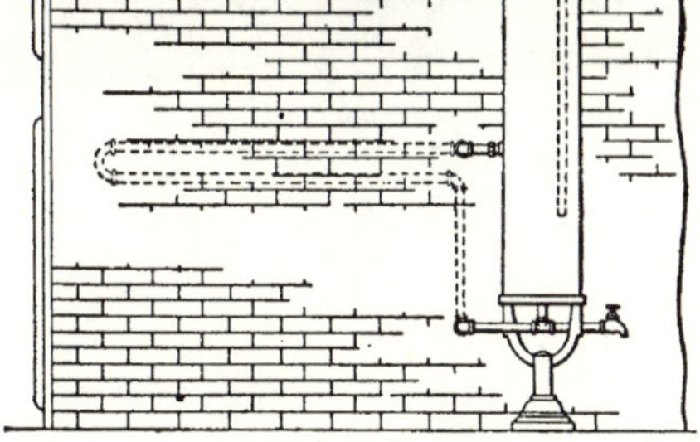

Fig. 125. A Bakery Boiler Heated by Coil in Oven.

CHAPTER XIII.

Air-Locking in Hot Water Supply Systems—Expansion of Water, Relief Pipes and Valves.

Much of the unsatisfactory service of hot water supply systems is undoubtedly due to partial air-locking in the pipes causing stoppage, or at least retarding of the circulation. This is especially evident in those systems supplied at low pressure from an overhead tank as then the conditions are more favorable to the holding back of water by a pocket of air formed by a slight depression in the line or by improper connections. When the tank is at some considerable elevation or the supply is taken directly from the city supply mains the trouble is not so apparent as then there is sufficient pressure behind it to overcome the resistance of the air in any trapped portion. Where lead is the material used in the construction the sagging that results through the expansion of the pipe and improper supporting of lateral runs is often responsible for an air pocket. Lead has a peculiar property that distinguishes it from the other metals used for water supply piping in that it "cannot come back." That is when it is expanded beyond its normal area or proportions it remains in the formation that the expansion has caused it to take. This is why lead pipe laid across joists without a supporting strip soon sags and forms pockets between each joist. The sagging is hastened and carried to a greater extent when the pipe is used for hot water and if the system is a circulating one the operation is very soon faulty. The same result may be experienced with the use of galvanized iron or brass pipe if it is not carefully laid so as to obtain a proper pitch. Where branch pipes to bathrooms and other apartments are carried up and a circulating pipe is returned either direct to the heater or to a return pipe serving other rooms as well, this fault may easily occur at the point where the return connection is made. The illustration in Fig. 126 shows what is meant and how the circulation is affected by the drooping of the branch pipe to the supply end either by insufficient support or by careless fitting. Another constructional cause of poor supply is shown in Fig. 127. In this case the complaint will be that no water can be drawn at the wash trays or that the supply is intermittent and poor. The cause is the trapping of air in the loop made by the supply pipe and indicated by the dotted lines in the illustration. It has already been explained that water at normal temperature contains a considerable proportion of air and that as the temperature of the water rises its capacity

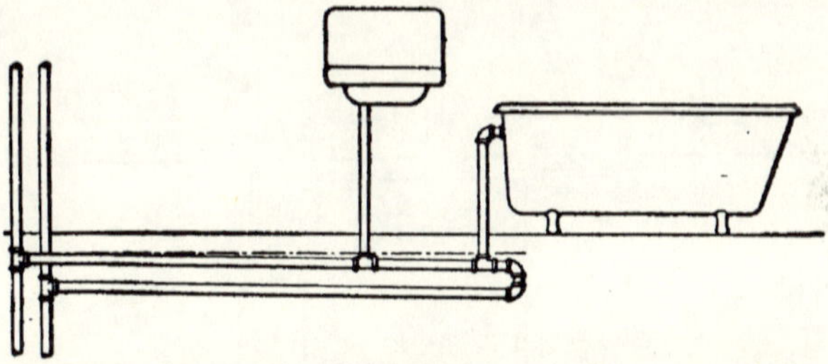

Fig. 126. Circulating Connections Inoperative Through Lack of Pitch.

to absorb air falls and therefore a considerable quantity is liberated when the water in the range boiler is brought to the high temperature that its heating in the water front gives it. This air has no means of escape and lodges in the highest point of the water supply system, forcing back the water as it does so. If then there is not sufficient head of water to overcome the resistance of this air pocket when the faucets at the wash trays are opened there will be no flow of water as the dip that is formed by the branch as it passes down to the basement forms a complete trap which prevents the escape of the air. Two methods of overcoming the trouble are shown in Figs. 128 and 129.

The first of these shows a circulating system with the return pipe taken off close to the tank and with the flow pipe continuing up to and turning over the top of the tank to allow of the escape of air as it is liberated. Another pipe in a similar position performs a like service for the cold water supply pipe and enables it to be emptied should the valve at the tank be closed. These pipes also prevent the accumulation of undue pressure in the system and obviate the expansion back into the tank of hot water from the boiler as would take place in the first arrangement. The second system shows an arrangement which has some advantage over the first in that it requires less pipe, the air in the branch pipe will be readily drawn off at the fixtures each time they are opened and there will be less chance of reversing the circulation by the act of drawing water. Another reason is that when the water in the tank is low there is a possibility of drawing a certain amount of air in through the expansion pipe owing to the slight resistance that the water above the return branch and the strong suction that the act of

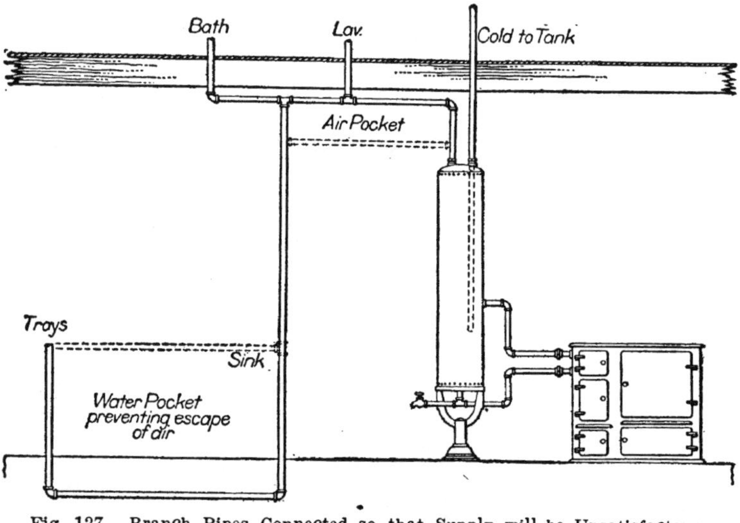

Fig. 127. Branch Pipes Connected so that Supply will be Unsatisfactory.

drawing it at a fixture on the lower floors has. This will cause a sputtering and intermittent flow at the fixtures. This disadvantage is avoided by using the construction shown in Fig. 129. This trouble is as likely to occur in large systems on the drop feed principle. When the supply to apartment houses and office buildings is laid out so that a circulation is to be maintained in each line and these are supplied from overhead as shown in Fig. 74 the utmost care must be taken in laying the lateral pipes at the top of the loop so that the large quantity of air which is being liberated in the system will have immediate vent. As a rule the house tank is at a sufficient height above the loop to offset any chance of drawing air back through the expansion pipe, but not at such a height that an air lock would be forced by the pressure of water that was available from it.

A Unique Remedy for Air-bound Hot-water Service.

In arranging for the hot-water service to a number of plumbing fixtures, as shown in the illustration, Fig. 130, the

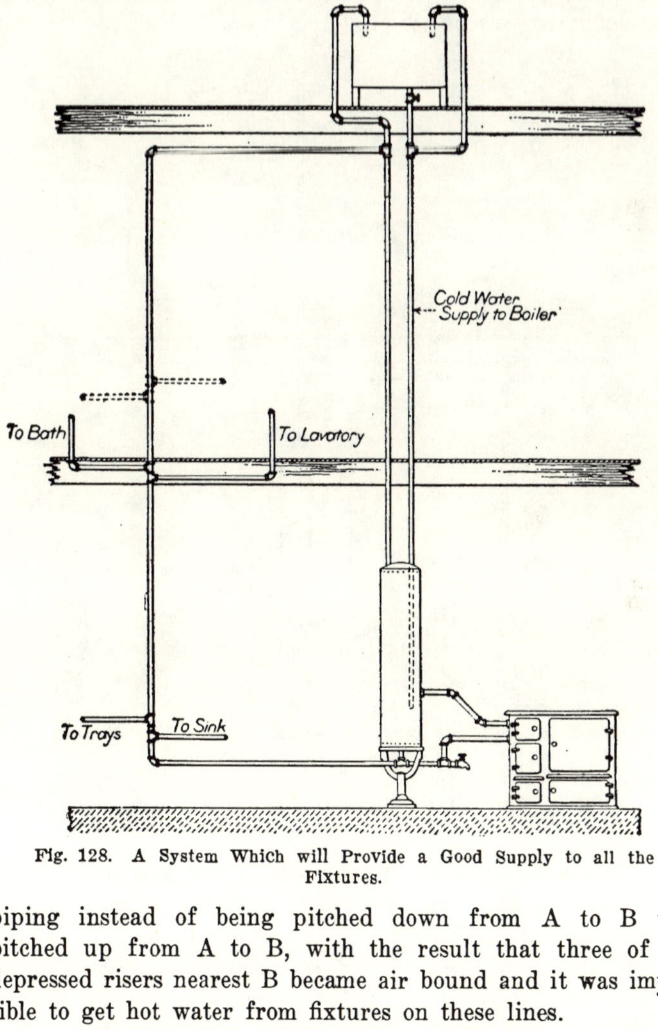

Fig. 128. A System Which will Provide a Good Supply to all the Fixtures.

piping instead of being pitched down from A to B was pitched up from A to B, with the result that three of the depressed risers nearest B became air bound and it was impossible to get hot water from fixtures on these lines.

An elevation of the piping system is presented in Fig. 130

with the tank heater and the hot-water storage tank at the left, showing a riser running directly up to the point A. The workmen were instructed to have the pipe to pitch down from A to the point B, but, unfortunately, this was not done. The piping was erected with a pitch in the other direction, so that the air which was carried into the system with the water from the street main was liberated at the top and gradually accumu-

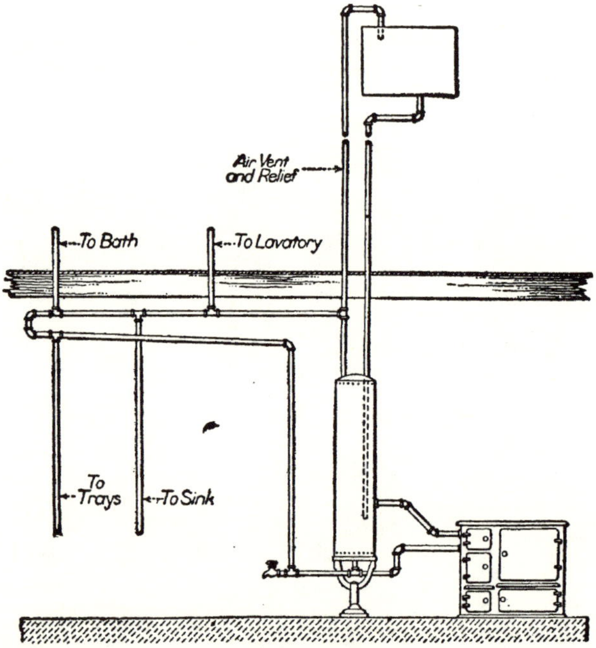

Fig. 129. Method of Connecting Branch Pipes to Ensure Proper Circulation and Supply.

lated to stop the circulation in the three supply risers at the right.

Naturally, complaints were made and it was necessary to make some change which would afford relief. It had been the expectation with the pitch in the other direction that whatever air was carried up and liberated at the top of the piping would be carried along in supplying the various fixtures and that no trouble would be experienced. An automatic air valve might have served to afford relief, but the question as to its continuing in good condition and operating continuously was too uncertain to permit its use.

As a result, an old boiler-feed regulator was connected at the point B and arranged as shown in Fig. 131. The ball cock was connected to the inlet in the ordinary manner. All of the other openings were plugged, except that at the bottom, which connected with the hot-water piping system. This allowed the

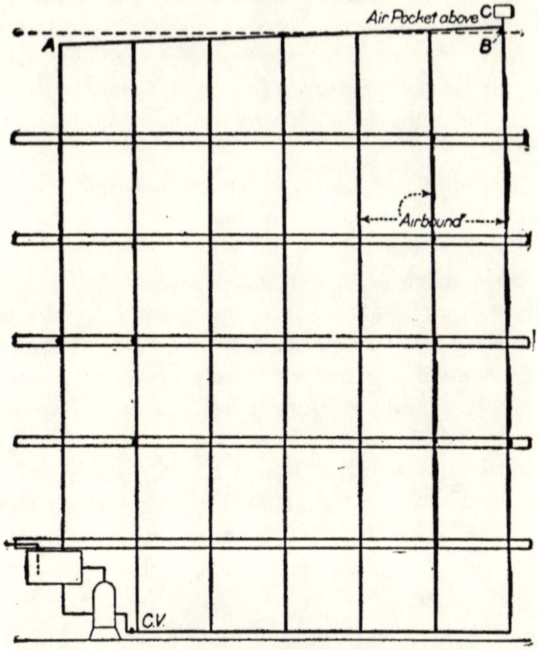

Fig. 130. Diagram Showing Air Vent Placed to Relieve System at High Point.

air to pass up the pipe into the tank and as it accumulated it forced the water down, so that the float opened the cock and allowed the air to escape. Thus it will be seen that when there was little water in the tank, the air was allowed to escape, but the float would rise with the water and close the cock to prevent the escape of water. A check valve was placed at the tank heater as shown

Expansion of Water Through Relief Pipes.

When a hot water installation is made with a supply coming from an overhead tank it is occasionally difficult to secure as much elevation for this as is desirable. This condition frequently obtains in houses of the bungalow type owing to the style of roof that this type calls for. When such difficulties arise and the tank is placed at a height very little above that of the highest fixture to be supplied or of the boiler as is shown in Fig. 132 there is likely to be trouble from too much hot water being forced back through the expansion pipe. This is of course taken to the tank and turned over the top with this purpose in view, but an excessive amount added to the water may increase the temperature of the water stored there so much that it will be unfit for drinking purposes and unpleasant to use for others.

The condition may arise in more than one way. It may be due solely to the increase in volume due to the heating of the water and its consequent relief through the expansion pipe. It may be due to collection of the steam and air bubbles liberated from the water in the water front by the process of heating, for it must be remembered that water at high temperatures will not hold as much air in solution as will water at normal temperature and therefore it is freed as the water becomes heated.

It may be due to circulation if the tank is low. This would be set up by reason of the expansion pipe virtually forming a flow pipe, the cold water supply pipe acting as a return pipe and thus causing the cold water tank to act as a storage reservoir for hot water.

To appreciate this the following table of increase of volume in the water at different temperature should be studied.

Increase in volume of water from 40 deg. to 250 deg. Fahr.:

Temperature.	Volume.	Pressure
40 degrees	1.0000	Atmospheric.
50 degrees	1.0004	"
60 degrees	1.0012	"
70 degrees	1.0023	
80 degrees	1.0038	
90 degrees	1.0055	
100 degrees	1.0074	
120 degrees	1.0121	
140 degrees	1.0175	
160 degrees	1.0238	
180 degrees	1.0307	
200 degrees	1.0384	
212 degrees	1.0433	"
230 degrees	1.0512	6 lb. pressure or over.
250 degrees	1.0604	15 lb. pressure or over.

This table will show that if the water is heated to only 180 deg. Fahr. the water will be increased in volume to some 1/30th of its bulk. That is if there is 60 gal. in the boiler and another 15 gal. in the pipes and water front there will be added to the tank in the attic by expansion from the water heated 2½ gallons of water.

This is when the tank is at such a height that the flow is not continuous through circulation as indicated or at which the water is forced back by air or steam pressure. To avoid these conditions the expansion pipe must be taken off the boiler in such a manner as to provide a free passageway for the air as it is liberated from the water and which will not tend to promote a circulation which will be maintained by the inflow of cold water to replace that sent up into the storage tank through the expansion pipe. The manner in which this can be done is shown in Fig. 132. If it is desired to keep all the water which is sent up the expansion pipe from mixing with the cold water in the tank it is a simple matter to fit an expansion tank proportioned to the size of the boiler and which will contain the hot water without allowing it to flow over into the storage tank. This is shown in Fig. 133. At the same time it must not be overlooked that a certain proportion of the increased volume will be relieved through the cold water supply pipe and that this will back up into the tank through the pipe. To guard against this a check valve may be fitted, but as the water will be cold anyhow this is not a great objection. An old fashioned method was to form a trap on the pipe supplying the cold water but this was generally done when the open type of hot water storage tank placed on the same level as the cold water cistern was used, the supply being taken from one tank into the bottom of the other. The trap prevented any circulation being set up between the two tanks, but did not prevent a certain

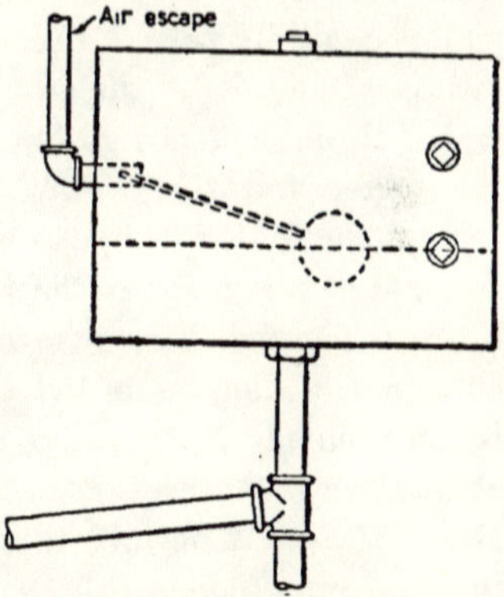

Fig. 131. An Automatic Air Relief Tank.

amount of the cooler water at the bottom of the hot water tank from backing into the other.

Safety and Vacuum Valves.

The provision of pressure relief valves is a safeguard which is usually adopted on all hot water tanks of any size and which could with advantage be used on ordinary kitchen range boilers in many cases. In towns where the use of water meters on the house supply is imperative and where the supply to the hot water boiler is taken directly from the main pipe there is no means of relief for the pressure generated by expansion of the water through heating should a check valve be placed between the boiler and the meter. This check valve is usually called for by the water company or municipal authorities to prevent hot water being forced back through the meter and so damaging it. The provision of a relief valve at the boiler obviates the chance of a dangerous pressure being raised when a strong fire is maintained in the range as it will open and allow a portion of the water to escape as soon as the pressure in the system rises to a point beyond that it is desired to carry. The valves in common use are of the ground seat variety with a spring which is adjusted to a tension equal to the pressure it is desired to carry in the system. Another type has a lever and weight which is adjustable to various pressures. These valves are shown in Fig. 134. Vacuum valves work in the opposite manner from safety valves, that is instead of opening when a pressure is raised they open when a partial vacuum is created. The object of using them is to prevent the contents of the boiler from being siphoned out through the supply pipe should the pressure in the cold water supply pipe fail for any reason. A break in the main pipe or the drawing of more than the ordinary amount of water from the main pipes such as might result from an excessive call on the main for fire purposes may result in the supply failing and as the water falls back from the boiler it creates a vacuum in the tank or pipes which might lower the water to a dangerous extent. Should this occur where a vacuum valve is fitted the vacuum would be destroyed by the admission of air through the valve which would open automatically. Combination pressure and vacuum valves may be had so that the danger of either excessive pressure or siphonage may be easily averted with a minimum degree of expense and fitting. When these valves are placed over a kitchen boiler they should be fitted as close to it as possible and if of the combination variety, should preferably be placed on the cold supply inlet, a tee being inserted for that purpose. If this is done practically no water will be siphoned out of the boiler when the vacuum valve comes into action as air is admitted

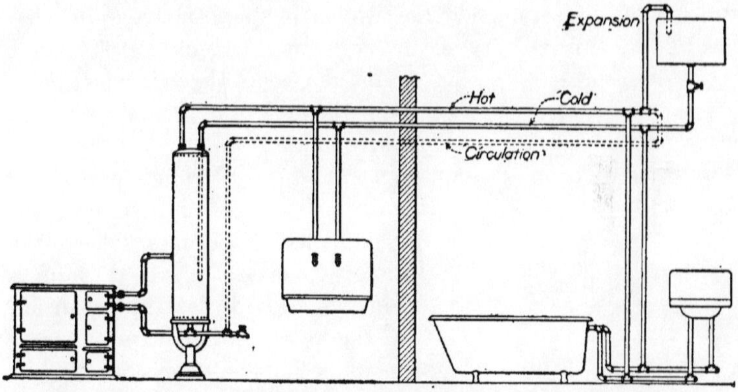

Fig. 132. Hot Water Supply System Liable to Give Trouble From Expansion of Water Into Tank.

directly to the cold water pipe. If placed on the other connection the water may be lowered to the level of the vent hole on the dip pipe before the vacuum is broken. Fig. 135 shows how the valve may be easily fitted and also how it should be fitted when the supply is taken into the tank at the lower side.

Collapse of Copper Boilers.

The use of a copper boiler of very light gauge is not advisable for two reasons. The first of course is the comparatively short life of the boiler through the extra wear and tear on it caused by repeated contraction and expansion which the thin shell is unable to withstand without giving out at some of the joints in time. The second is the liability to collapse due to accidental formation of a vacuum in the boiler by siphonage or condensation of steam formed therein by overheating. When a vacuum is created the thin shell is unable to resist the atmospheric pressure of 14.7 lb. per sq. in. and the boiler collapses. So long as the boiler is full of water this pressure has no effect on it of course or even a much heavier pressure will be withstood if the pressure is on the internal surface as is the case when the supply comes from a city main. When a vacuum has been created and the boiler has been crushed it is not uncommon for the returning water to expand it again so that it regains its original shape, but this of course is exceptional and the correct course is to make provision for the absolute prevention of a vacuum. If the system is supplied from an overhead tank this is very easily done by carrying a relief pipe from the top of the boiler to a point over the tank or by fitting a vacuum valve at the boiler.

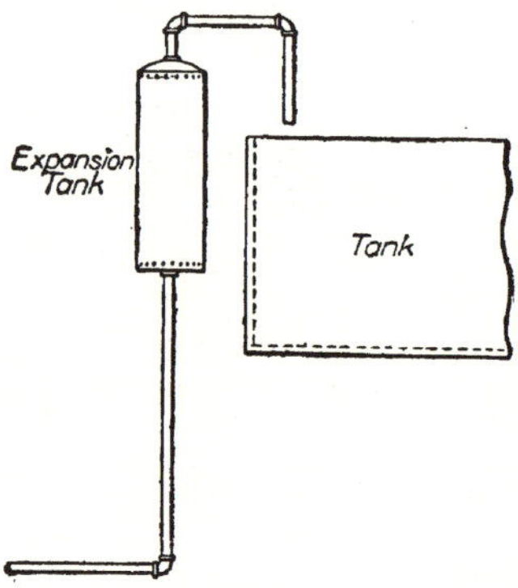

Fig 133. Method of Connecting Expansion Tank to Avoid Flow of Hot Water Into Storage Tank.

The use of relief pipes and vacuum valves is not general, but is adopted sometimes where the supply pressure is low or from a tank, to provide for the expansion of the water or the escape of air or steam. They also permit air to enter the boiler to prevent the formation of a vacuum when the boiler is emptied of water by siphonage or by the condensation of steam. The proper place to connect the valve or relief pipe is at the top of the boiler, but if the hot water pipe rises direct from the top of the boiler without any dip the relief pipe may be connected at the highest point and run up above the level of the water supply.

Where the supply to the boiler comes directly from the street main the vacuum valve effectively prevents the boiler from being siphoned empty should a break occur on the main or other cause operate to allow the water supply to fall back in the pipes. As soon as the pressure is removed and the water falls back drawing air behind it the vacuum valve opens and effectively prevents the water from being forced out through the supply pipe by atmospheric pressure. The usual provision of a vent hole in the boiler tube is not an absolute safeguard as it is not operative unless a faucet is opened to admit air and then only when the water falls below the small hole. As this is generally kept down from the top of the boiler so that cold water will not pass through it too quickly and mix with the hot water leaving the boiler this means that at least six inches of the water in the boiler will be siphoned before the action ceases. If then the faucet should be closed and the water in the system become overheated steam forms in the boiler and on the return of the cold water its sudden condensation may form the vacuum and the boiler collapse. Copper boilers should be made of heavy sheet and in addition should be reinforced with bands brazed on the internal surface. This will give them stability to resist atmospheric pressure at least. Fig. 136 shows an installation in which the boiler collapsed and the conditions which led to it.

It is certain that the collapse was caused by steam in the boiler due to great heating capacity of the pipe water front

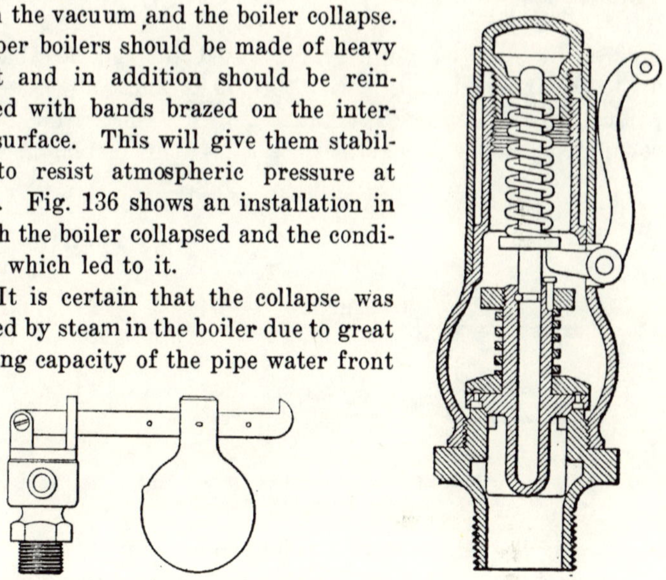

Fig. 134. Two Types of Relief Valves.

and lack of pressure in the supply. From the top of the boiler to the level of the water in the tank, as shown in the illustration, would hardly be over 6 feet. Using the thumb rule of water pressure of ½ pound to each foot in height, a pressure of only 3 pounds would be exerted at the top of the boiler by the supply. The large water heating surface exposed by the construction of the water front would enable steam to be generated freely and passed to the boiler. The accumulation of steam at the top of the boiler would drive the light pressure water supply back to the tank and would discolor the copper, slightly. Drawing off hot water at any faucet would let cooler water enter the boiler and pass to the water back, when the generation of steam would be stopped and the steam from the boiler would follow along the hot water service pipe, both actions tending to condense the steam and create a vacuum. If the vacuum was of considerable extent it would not be filled with water before the atmospheric pressure of about 14.7 pounds to the square inch on the surface of the light copper boiler would cave it in.

A repetition of the conditions causing the collapse was rendered impossible by taking a relief pipe from the outlet pipe at the boiler as shown by the dotted line and carrying this up to the tank, turning it over the top to allow any water expanded up through it to fall back into the storage tank. The collapse of double boilers may be caused by making the con-

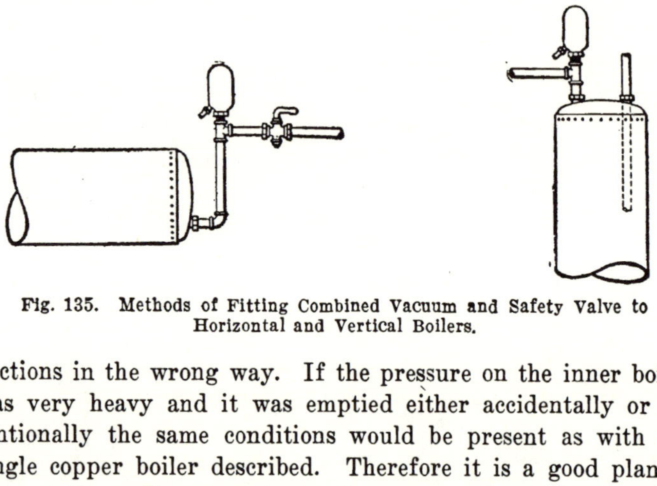

Fig. 135. Methods of Fitting Combined Vacuum and Safety Valve to Horizontal and Vertical Boilers.

nections in the wrong way. If the pressure on the inner boiler was very heavy and it was emptied either accidentally or intentionally the same conditions would be present as with the single copper boiler described. Therefore it is a good plan to connect the sediment cocks in the manner shown in Fig. 137 so that the outer boiler must be emptied before the contents of the inner one can be drawn off. This is the practice where there is any possibility of collapse occurring and as it is no more difficult to connect in this fashion than in any other the safeguard is worth making in all cases.

Method of Avoiding Excessive Pressure in Hot Water System Where Check Valves Are Used.

In most water supply systems where meters are required by the water company or municipal authorities it is obligatory to fit a check valve on the main pipe at a point behind the meter.

This is intended to prevent the hot water as it expands from backing into the meter and so damaging the parts with which it would come in contact. While this effectually prevents the damage to the meter it leads to a dangerous condition in the hot water system as when the boiler supply is taken directly from the main pipe there is no means for relief of the water

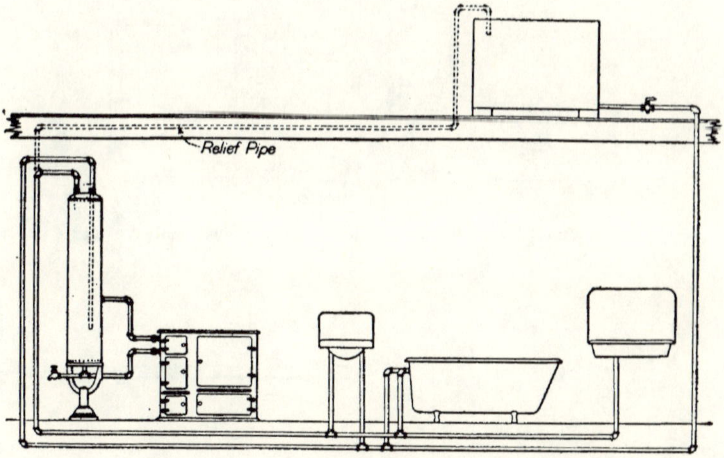

Fig. 136. A System in Which the Boiler Collapsed and the Means Taken to Avoid its Recurrence.

when its volume is increased by heating and a dangerous pressure is liable to be raised.

This danger can be avoided by the provision of a safety valve on the boiler, but there is another method which appeals to many who do not care to have the safety valve. This consists of making a by-pass around the meter and check valve and in this by-pass to fit another check valve with the swinging valve working in the opposite direction. This as will be seen from the illustration in Fig. 138 allows any excess pressure generated by the expansion of the water to be relieved through this valve, as immediately the pressure exceeds that carried in the water main the check valve will open and allow water to flow back into the main pipe.

Either a globe or a swinging check valve may be used but the latter is generally used in hot water work owing to its lighter action. Another method is simply to drill a small hole in the check valve that is fitted behind the meter. This is calculated to relieve the pressure as it is generated but not to

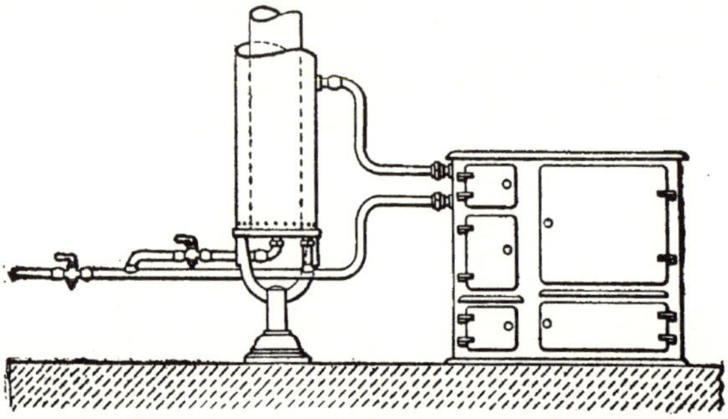

Fig. 137. Method of Connecting a Double Boiler.

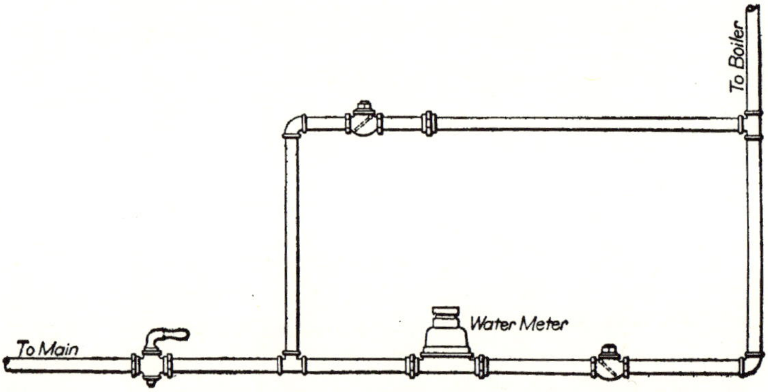

Fig. 138. Use of Check Valves and By-pass to Prevent Undue Pressure From Expansion.

pass any more than a tiny stream of hot water, which mixing with the cold water in the pipe would not appreciably raise its temperature.

CHAPTER XIV.

Common Complaints and Their Remedy—Unsatisfactory Heating of Water.

Occasionally a complaint will be received that no hot water can be drawn at the fixtures and that even when a good fire is maintained in the range the trouble continues. There are several causes which may operate to cause these conditions, but the most common probably is that the fire is not maintained in as good condition as the people believe. On examination of the fire box when such a complaint is made it is common to find the fuel in contact with the water front burning very dull, and the cause of this may be that the fuel is banked up too high in the fire box under the mistaken notion that the heavy firing will add to the heating power of the boiler. As a rule it simply obstructs the passage of the hot gases to the flues and prevents proper combustion. The failure to keep the fire clean will of course have the same result as the supply of oxygen is then insufficient to support proper combustion and there is neither satisfaction or economy in maintaining a fire in this condition. Then the size of the coal may be the seat of the trouble. If the grate and the provision for the admission of air is not designed to suit coal of the smaller sizes a bright and hot fire will be impossible when it is used and conversely, if too large coal is used the pieces do not lie close enough to the heating surface to have the full effect that is necessary to heat the water properly.

In many modern ranges also the fire box is so narrow that the heat transferred to the water front cools off the coals to such an extent that the layer in contact with the heating surface is not consumed at the same rate as the rest. With a deep fire box the heat in the general body of the fire is sufficient to overcome this cooling effect without affecting seriously the value in heating the other parts of the stove. This skimping of size is responsible for much of the trouble that is experienced when water fronts are put into stoves that previously were used for baking only, and where the difference in the baking qualities of the range after the installation of the water front has been very noticeable. In such cases it is better to use a coil of brass pipe than the water front as its cooling effect is much less noticeable.

Stoppages in Water Front May Cause It.

Another cause of bad service is that occasionally parts of the core, pieces of wire and sand are left in the water front and these lodge in the passageway at the end, which connects the upper and lower parts of the water front. This may be present in quantities sufficient to seriously affect the circulation to the boiler and yet not in sufficient quantity to stop it altogether and so make its presence known by snapping sounds due to formation of steam in the water front. In such a case very careful inspection is necessary to locate the stoppage and to see that it is entirely removed. The other things to look to are the arrangement and size of the connecting pipes to the boiler, the amount of pitch given the pipes, the number of elbows on the connections and the distance of the boiler from the stove. If this latter is too much it is quite possible that the water is cooled to such an extent before reaching the boiler that the circulation is very sluggish. If the sizes of the pipes and connections are too small, of course enough water is not being circulated to keep the contents of the boiler up to the temperature desired. The same thing applies to the pitch of the pipes and the number of elbows on it. There may be an unnecessary amount of friction to overcome.

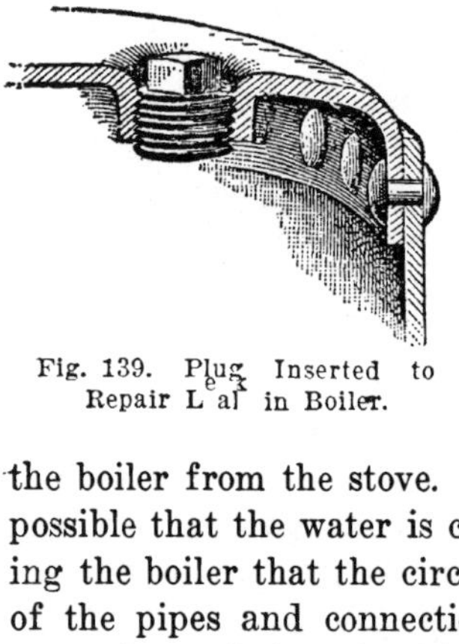

Fig. 139. Plug Inserted to Repair Leak in Boiler.

The Boiler Supply Pipe May Be Rusted Off.

Still another reason for the poor supply may be that the dip pipe on the cold supply to the tank may have rusted off or the connections may have been crossed and the hot supply pipe been connected to the cold by accident. Again the water front may be encrusted with lime and so practically insulated. If the supply system is a circulating one it may be that water is being drawn through the return pipe to the fixtures instead of by the regular flow pipe, and that a light check valve will have to be fitted to prevent this reversal when a faucet is opened. These are the factors which generally operate to cause trouble and which should be tracked down when investigating a complaint.

Corrosion of Kitchen Boilers and Temporary Repairs.

When a galvanized iron range boiler begins to show signs of corrosion after years of service this generally appears in the form of small pinholes, the major portion of which will probably be in the upper part of the boiler. Repair plugs are made which are easily inserted in such holes, these being made of steel, tempered, and of such a shape that they form for themselves a thread on the shell of the boiler by the action of screwing them into place. A shoulder on the plug is fitted with a fiber washer, and this on being made up against the boiler shell secures a water tight joint. Should the corrosion have progressed too far to use such a small plug it may be possible to repair the leak by driving into the hole a tapered steel pin which will turn the metal inward and so form a surface which may be tapped for a standard pipe size plug. Fig. 139 shows how this may be accomplished. Another way of stopping a leak of this nature is to drill out the corroded part securing a hole about 3/8 in. dia. Then a hexagonal headed machine bolt about 3/4 in. long and about 3/16 in. thick is secured and a nut fitted to it. A brass washer about 7/8 in. diameter which will fit snugly over the bolt is then slipped over it and a well fitting fiber washer placed behind it. The method of fitting this plug is to take out one of the boiler unions in the crown and then to lower the bolt into the boiler by means of a thin string which is wound around it, and which hangs down below the bolt. A piece of wire bent in the

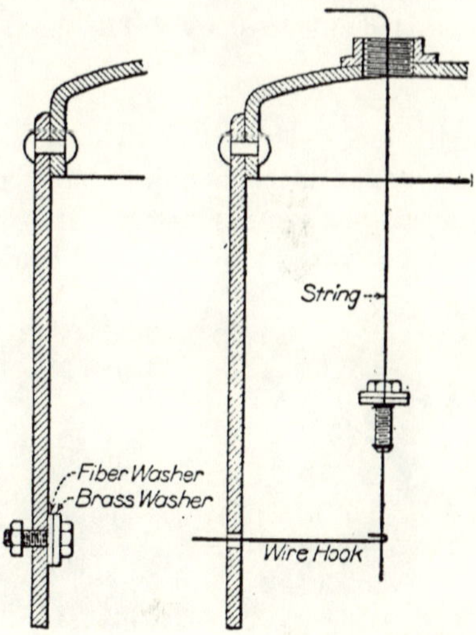

Figs. 140 and 141. Method of Inserting Bolt and Washer to Repair Leak.

form of a hook inserted through the hole which has been drilled to enlarge that caused by the corrosion is made to catch the string, and so pull the bolt through the hole, as shown in Fig. 141. Then the nut is put on and the washer drawn tightly up against the inside wall of the boiler. Before putting it into place the shell should be scraped free of rust by a bent wire filed so that it has an edge which will remove the unevenness of the corroded surfaces. Fig. 140 shows the bolt with its washers and nuts.

Hot Water Has a Milky Appearance.

A not uncommon complaint is that water as drawn from the faucets has a milky appearance, which disappears in time, leaving the water as clear as that drawn from the cold water faucet.

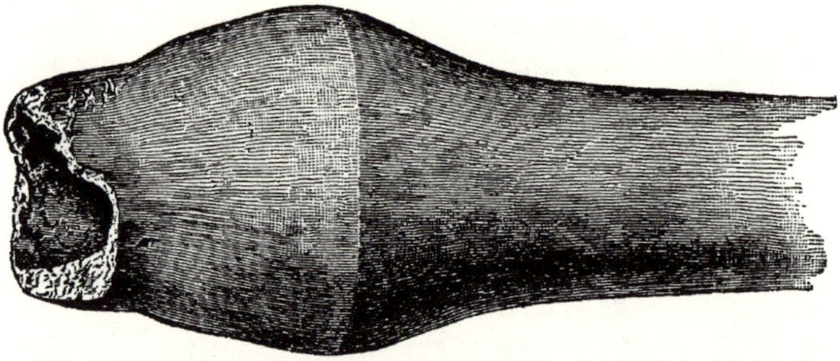

Fig. 142. A Wiped Joint Swelled and Broken by Water Hammer.

In cold water supply pipes this condition is sometimes noticeable in systems where a pump is used to elevate the water. It is also noticeable in the supply to houses on elevated situations on a city supply, and is due to air becoming emulsed with the water when more than the water will absorb at its normal pressure and temperature is present. This air may have been collected in the system when the pipe was partly emptied for repairs, and as the water was turned on again it is driven before the water and collects at the highest points, where it escapes with the water, giving it the appearance noted above. In the case of hot water the milky appearance is due to the water being full of small globules of steam. This condition is seldom found except with heavy firing or with a large water back when but little water has been used for a time. This enables all the water in the boiler to be at very nearly an even temperature. The boiler being subject to a pressure from the city mains, sometimes as much as 80 pounds, enables the water to be heated to several degrees above the boiling point, and immediately on opening the faucet this pressure is very materially reduced, enabling the water in the boiler and piping to expand into steam. It runs in this condition with the milky appearance mentioned, and as all of the steam cannot escape instantly some will be carried to the vessel, and as it gradually passes off the water will become clear. If the water is what is known as hard and more or less impregnated with lime in solution the excessive heating will liberate the lime, which, if in sufficient quantity, will give the effect described but which will disappear as the lime settles.

Water Hammer in Boiler Connection.

The subject of the illustration, Fig. 142, was furnished by a Philadelphia plumber, and represents a piece of pipe which was taken from a job where he was called to make some repairs. The broken end shows the solder in a joint that was wiped on the side connection to a kitchen boiler, and which was swelled and broken, it was thought, by water hammer. The enlargement is very uniform, there being no special swelling in one side, as is usually the case, and it will be noticed that the solder has stretched quite as much as the pipe. The pipe is very heavy, ¾-inch pipe, which shows the power of the concussions which finally broke the joint, after a leak which occurred in the swelled part had been stopped by hammering metal into the opening. To those who are not acquainted with water hammer it is explained as being due to steam that has formed in the water back passing to a point where it condenses, creating a vacuum which is filled by an inrush of water that strikes with a much greater force than is generally appreciated. When a water back is large and steam is generated freely this striking is frequent, and the result is as shown. A larger opening or a water way of the full size of the pipe through the side connection would carry the steam into the boiler before it condensed.

Removing a Lukewarm Water Complaint.

Whether there is any justice in it or not the stove manufacturer is called upon to make good many things for which he is in no way responsible, and a short time ago it was necessary for a manufacturer to send a salesman some distance to look into the cause for a faulty service in the hot-water supply from one of his ranges. Fig. 143 shows the manner in which it was connected. As can be seen, there is nothing wrong with the connection between the range and the boiler. This range and the water front are frequently used in connection with boilers of larger capacity, and have ample capacity for the 30-gal. boiler with which it was connected in this case. Nevertheless, the range was blamed for the trouble, as is the usual experience of stove manufacturers, until it was demonstrated that in place of the range and water front being at fault the trouble was due to the gas water heater connection, which though it had a stopcock to shut off the circulation through the water heater when it was not in use, was not turned off. In consequence, it was a simple matter for the cold water flowing into the boiler whenever the hot water faucet of any fixture was open to short-circuit through the bottom of the boiler and up through the gas heater into the hot-water service pipe to mingle with the hot water and give only a lukewarm supply of water. Just as soon as this stop-cock on the supply pipe to the gas water heater was shut off the trouble immediately stopped.

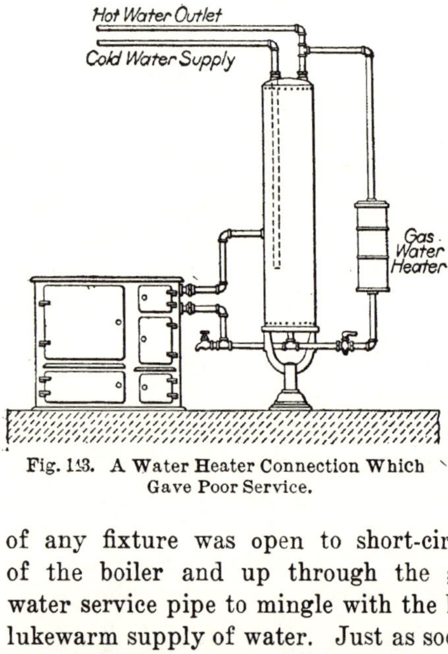

Fig. 143. A Water Heater Connection Which Gave Poor Service.

Apparently this customer had had similar troubles in a number of cases, but had never before learned exactly how to get over them, and said it was worth a great deal to him. It certainly entailed an unnecessary expense upon the manufacturer, but fortunately they are the exceptions, and in most instances the men to whom furnaces, ranges and stoves are sold know their business so thoroughly well that they do not overlook just such troubles as this and solve their problems alone. In consequence there are no hard feelings when occasionally a good customer finds himself stumped and needs assistance.

Cold Water Drawn at Hot Water Faucet.

In a system of hot water supply where a boiler on the second floor was heated by a coil in the basement and one in the kitchen range, there was a complaint that it was impossible to draw as much hot water as should be expected. The connections were made as shown in Fig. 144, and it was found that when the faucets were opened the cold water rushing into the boiler at the same end as that at which the supply pipe was connected short circuited, and instead of forcing the hot water out of the boiler to the faucets, took the shortest path through it and only a small quantity of hot water was sent into the pipes before the cold water followed it. The trouble was remedied by making the connections as shown in Fig. 145. In this it will be seen that the cold water supply enters the boiler at the opposite end to that at which the hot water leaves it, and this is sufficient to force all the hot water it contains out as the cold water enters. The trouble could have been remedied by using dip pipes inside the boiler also, but this is the most certain way.

Rusty Water.

One of the chief causes of complaint, and one that is very hard to eradicate, is that of rusty water. The rust may be proceeding from one of many causes, and it may be impossible to prevent its accumulation in the boiler, as for instance when the water that is supplied had a particularly active effect on the pipes and water front. There is, however, a method of making the connection that will alleviate the trouble greatly. If the boiler supply tube be taken out and a brass tube substituted which will have the end plugged and a number of holes drilled around the pipe at its lower extremity it will be found that the rush of water into the boiler instead of stirring up the rust on the bottom will be directed against the sides, and will therefore have no such effect. The conditions can be still further improved by the use of a duplex boiler connection, as shown in Fig. 146, for the hot water circulating connection, as in this case the water entering the boiler will also be delivered into the body of water already in the boiler, and so avoid any eddies while the hot water is drawn from the highest point in the boiler. The boiler tube drilled and plugged as described is shown in Fig. 147.

Making an Extra Connection to a Boiler.

For the purpose of making an extra connection to a boiler to suit the needs of a circulating system of supply to a distant

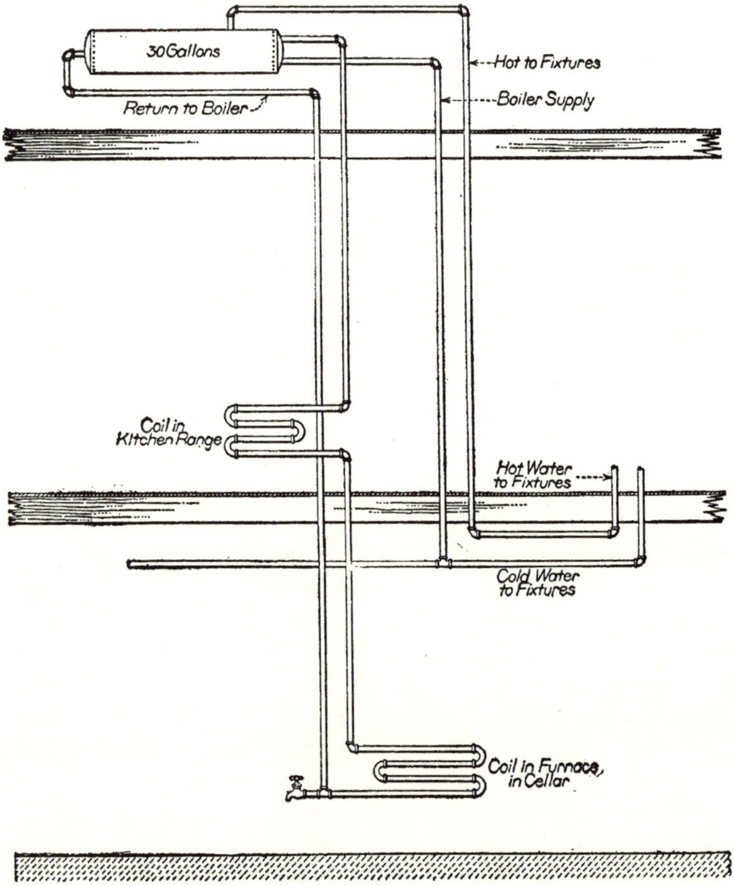

Fig. 144. Connections to a Horizontal Boiler Which Caused Short Circuiting.

fixture or to take in the return from a radiator which it is desired to heat from the kitchen range, it is sometimes necessary to drill a hole and make a new connection to the boiler. Fig. 148 and the description shows how this may be easily and effectively done. After the water has been withdrawn from the boiler it is a comparatively simple matter to arrange for another connection with the boiler shell. First, it is necessary to make a hole in the side at the desired point and of the right diameter, to allow the piece of threaded pipe, preferably 1 in. in diameter, to be in-

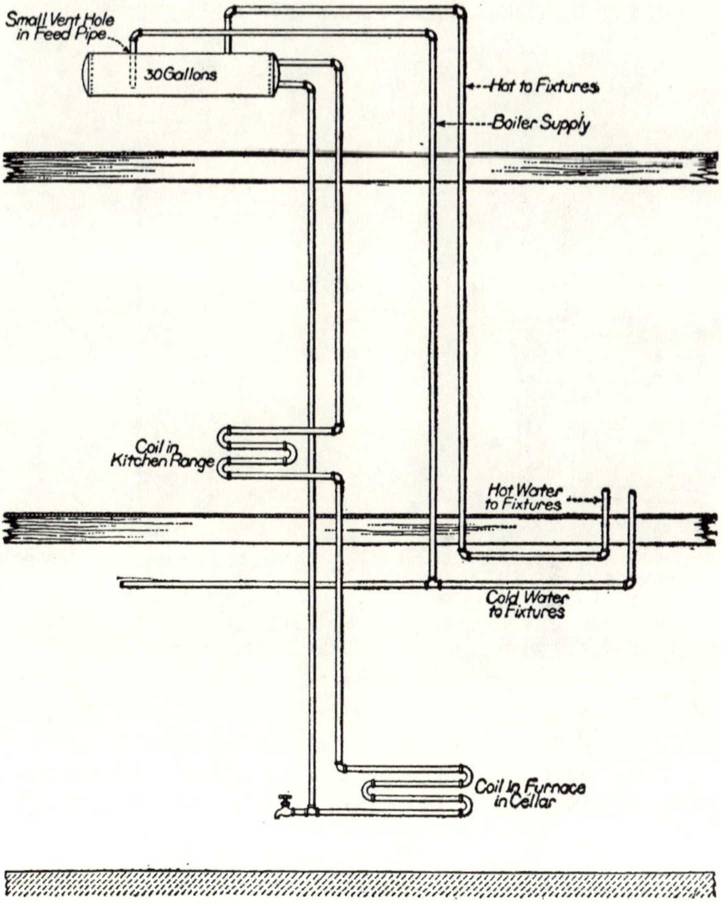

Fig. 145. Changes Made in Boiler Connections to Provide Satisfactory Supply.

serted. Next it is necessary to file at opposite points on the circle notches just wide enough and deep enough to allow a lock nut to be passed through. If a hole is drilled in the lock nut so that a wire can be inserted for holding it until the pipe is screwed into it, it will facilitate the work of making the connection. The pipe should have a long thread, so that a lock nut can be screwed up against an iron washer covering a soft washer, all of the threads on both washers and the surface of the boiler being coated first with rather thick red lead. Then when the lock nut is screwed up tight the opening in the boiler will be effectively cov-

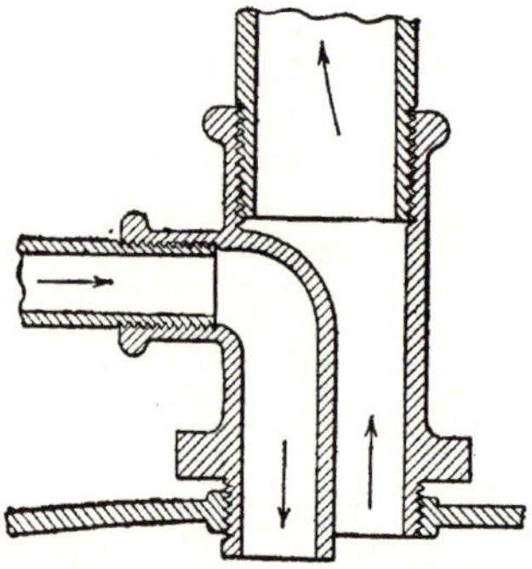

Fig. 146. A Duplex Boiler Connection Which Prevents Short Circuiting.

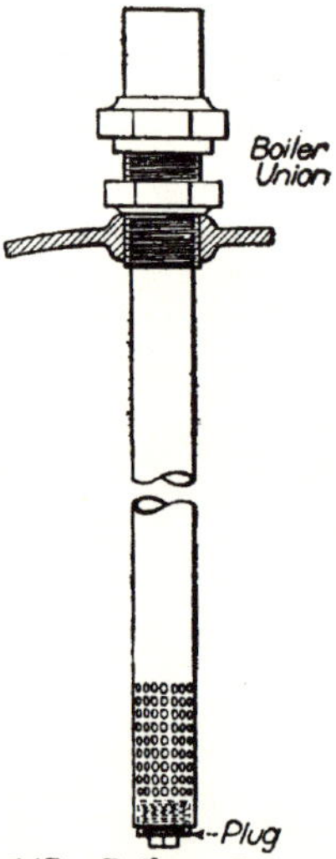

Fig. 147. Boiler Tube Which Will Not Stir Up the Sediment.

ered, as shown in the illustration herewith. Owing to the boiler shell being curved, it is well to have the soft washer pretty thick, or to form the iron washer covering it to the shape of the boiler shell, and when the lock nut is screwed up some strain should be put on the wrench to make sure the joint is tight. The outer end of the nipple is then ready for any pipe connection that may be needed.

Hot Water Supply to Barber Shop.

A satisfactory outfit for the supply of hot water to a barber shop may be made as described in Fig. 149. This consists of a copper tank 13½ in. diameter and 13 in. high made out of 16 oz. copper, so that it will be stiff without the use of beads. The top is made in the form of a flat cone in the center of which is left a hand hole with a cover 5 in. diameter. This is made to lift off, so that the tank may be filled with water as required. Immersed in the water is a coil of ½ in. brass pipe made by winding the pipe around a wooden block until turns enough have been made. The coil is 8 in. in diameter and contains 12 to 18 ft. of pipe as may be required. Eighteen feet of pipe will heat enough water to supply a shop with four chairs even when they are busy. The coil is fastened into the tank by lock nuts, a gasket being put between them and the copper.

The burner chamber is made of black sheet iron with 1 in. holes punched around the upper and lower edges. The chamber is 7 in. high and a hole about 4 in. by 6 in. is cut to allow of the burner being lighted. It is better to leave this without a door, as there is more air then for proper combustion. An ordinary Bunsen burner can be used or a small boiling ring such as is used for cooking. The space between the base and the stand on which it is placed may be filled with sand to guard against fire. The apparatus, of course, is connected with the water supply system and provides hot water instantaneously, so a valve is fitted to control the supply and to pass only as much water as can be heated to the desired temperature. The copper chamber requires attention, and must be kept filled with water, but if the cover is a good fit there is not much loss by evaporation.

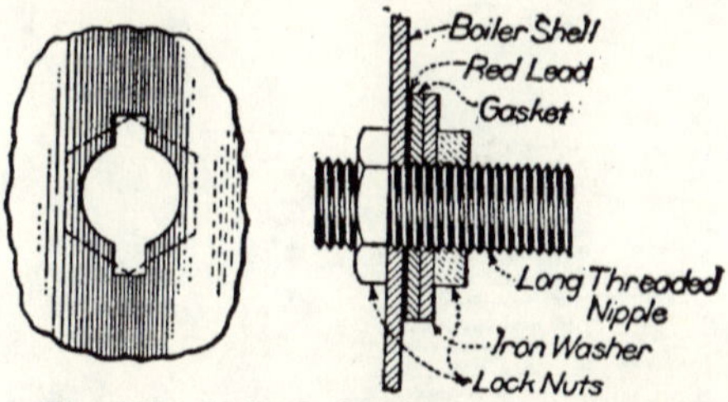

Fig. 148. Method of Making Extra Connection to Boiler.

Comparative Value of Lead and Brass for Range Connections.

Lead pipe as a material for the connection of range boilers and water fronts possesses only one recommendation—its freedom from corrosion. In many towns the use of lead pipe is almost compulsory owing to the rate of corrosion of iron pipe with the water provided for public use, and in these towns there is generally quite a large repair business done in connection with the replacement of range boiler connecting pipes or the repair of leaks at joints. This is due to the great amount of wear and tear caused by the expansion and contraction of the pipes, and as lead does not withstand this movement very well it soon gives out at the bends or joints where the motion is retarded. Sagging of the pipes also occurs. This is due to the pipes becoming very easily bent when heated and by the expansion lengthening them and the inability of the lead to recover its original shape entirely when cooled again. The connections to the boilers and water fronts are made by means of brass unions wiped to the lead pipe and screwed into the iron. The method of connecting a boiler with lead pipe is shown in Fig. 150. Brass pipe may be used for boiler connections in the same manner as iron; that is, elbows, tees and other fittings may be procured in the same designs and in the same sizes as in iron. It may also be easily bent, and when this is done neatly it offers a medium for a permanent and smooth working job, as there are no sharp turns to obstruct the circulation. When brass pipe is threaded and screwed into a fitting of the same material care should be taken not to force the thread in further than is necessary to secure a tight joint. If a thread with much taper is cut, and it is attempted to screw the thread far enough home to hide all the thread it may easily happen that the fitting will be spread and it will be impossible to secure a tight joint. It is better to see that a good sharp thread is cut, and that it is not so full that the whole length cannot be made into the fitting without straining it severely. It is advisable to examine all fittings for sandholes before inserting them also, as they are liable to contain small flaws which are easily overlooked until the water is turned in. Brass pipe does not corrode like iron, and there are no minerals met with in potable waters that have any injurious effect on it, although it is of course liable to become stopped with sediment like other pipes. A method of connecting a range with brass pipe

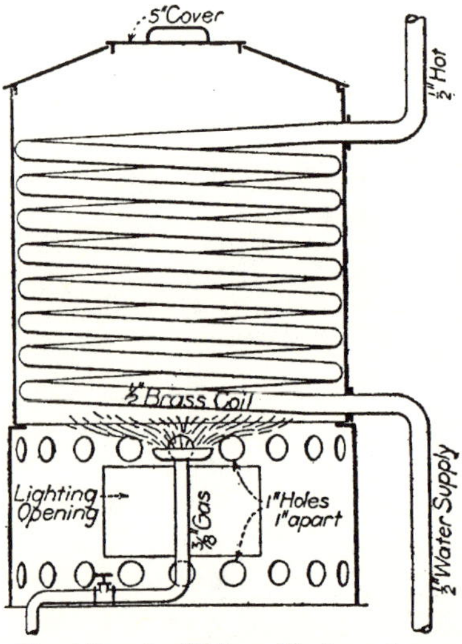

Fig. 149. A Water Heater for a Barber Shop.

showing the unions used, and also supports for the gas heater, is shown in Fig. 151. The nipple between the cross tee and the floor flange is made solid, this feature being intended only to secure rigidity. It is customary to use unions at the water-front end and it is often more convenient to fit them at the boiler as well. This applies especially where the pieces are long and where it would be difficult to turn the piece into the boiler tapping direct. In cutting threads on brass pipe some men will

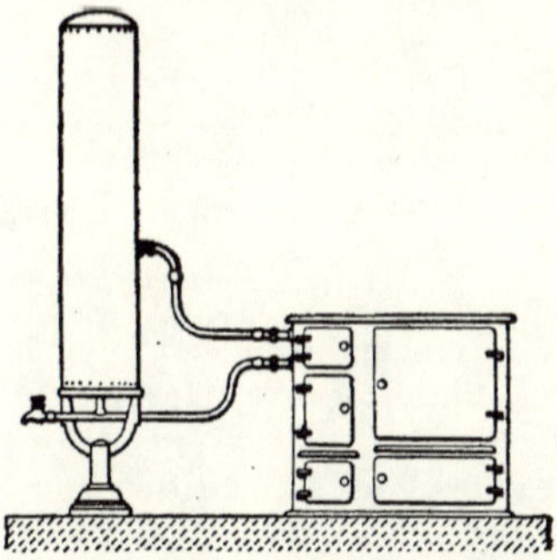

Fig. 150. Range Boiler Connected With Lead Pipe.

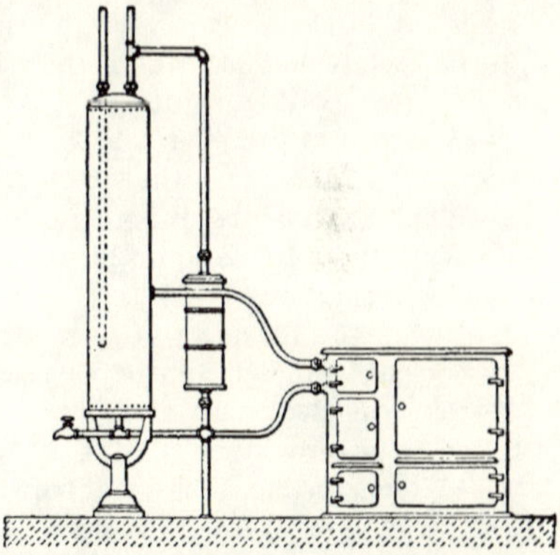

Fig. 151. Range Boiler Connected With Brass Pipe.

not allow oil to be used, but the point is a minor one, and if a good grade of lubricating oil is employed the die seems to work better when treated as if iron pipe were being threaded.

Copper pipe is used to a great extent in English practice, but has no advantage that brass does not possess. It is imperative that it be tinned, inside at least, and it is commonly supplied tinned outside as well. Two grades are procurable. Hard copper pipe is best suited for work where fittings are employed, as it will stand up as well as brass or iron pipe. Where it is desired to bend the pipes, soft or annealed copper is preferable. This is easily worked, but requires hangers at frequent intervals when fitted in long lengths. Brass pipe may also be supplied in annealed or semi-annealed form and is easier to bend for boiler connections than the hard brass quality.

TYPICAL EXAMINATION QUESTIONS ON THE THEORY AND PRACTICE OF HOT WATER SUPPLY INSTALLATION.

CHAPTER 1.

Pages 7 to 15.

1—Define the meaning of the word "Circulation" as applied in the practice of hot water supply fitting.

2—What is the cause of circulation?

3—Is the density of water the same at all temperatures between freezing and boiling points?

4—What takes place in a body of water when heat is applied to the lower part of the vessel in which it is contained?

5—What do you understand by the term "Motive Column"?

6—Why should a dip in the pipes of a hot water supply system prevent circulation?

7—Define the meaning of Combustion.

8—What is the relation of Heat to Combustion?

9—Can you define an application of radiation, convection and conduction in the practice of heating water?

10—Which means of transmitting heat is the most effective, radiation, convection or conduction?

11—Why does the production of smoke and its emission at the chimney of a heating plant indicate a lack of economy?

12—What is necessary in a firebox to secure the highest economy and most perfect combustion?

14—What is the average heating power of anthracite and the average amount of heat per pound of fuel transmitted to the water?

15—What is the difference between a strain and a stress?

16—What is the relation of elasticity to elastic limit?

CHAPTER 2.

Pages 16 to 24.

17—Is a pipe conveying hot water more liable to corrosion than one conveying cold water?

18—What is the difference between corrosion and sedimentation?

19—What are the principal causes of deposits in water fronts and boilers?

20—Why should corrosive influences appear to be most active at temperatures between 140 and 180 Fahr.?

21—In view of the more severe corrosive influences in hot water systems operating at high temperatures, what may be done to mitigate the trouble?

22—How is the hardness of water measured and classified?

23—Can a permanently hard water be used without danger of precipitating sediment in the water front or pipes?

24—How may the presence of lime in water be made evident in a hot water supply system?

25—What is a good method of removing lime and kindred deposits from water fronts and pipes?

26—Is there any method of piping a hot water boiler to avoid precipitation of lime?

27—What is the best type of fittings to use in pipes liable to stoppage by such deposits?

28—Is there any appliance that may be fitted to the connecting pipes of a boiler and water front to facilitate removal of sediment?

29—In what conditions is it important that there should be no heat loss by radiation from hot water pipes or storage boilers?

30—Can you describe any method of preventing heat losses?

31—What effect on the rate of heat loss by radiation has the painting of a storage tank?

32—Is the heat loss the same with the use of all kinds of paints if the same number of coats are applied?

CHAPTER 3.

Pages 17 to 33.

33—What are the correct positions of the tappings for connections in water fronts or water backs?

34—What is the object of having a partition in a water back?

35—Is there any danger of damage being done to water fronts by accumulation of sediment or steam?

36—What indication is given of accumulation of sediment in the appearance of a water front?

37—What is the most effective position in the firebox for a water heater?

38—How may the water heating and baking qualities of a range be depreciated by the proportions of the water front and firebox?

39—What is about the maximum exposure of heating surface possible in a kitchen range water back?

40—Why should water fronts be bedded with cement or fireclay?

41—Is it important that the stove should be set level?

42—How may the heating surface of a water front be easily extended?

43—How are coils for water heating made to fit kitchen range fire boxes?

44—What should be particularly guarded against in fitting coils of the usual type to a kitchen range?

45—Is the type of coil fitted in a position over the fire box as efficient as that which is in contact with the fuel?

46—How is the capacity of a water front computed?

47—What is the highest rate of heat transmission that should be estimated in water fronts and coils in proportioning sizes of a water supply?

48—Should the maximum requirements of a household be made the basis on which to estimate the proportions of heating power required?

CHAPTER 4.

Pages 34 to 45.

49—What is meant by "allowing a swing" to the joints in a range boiler connection?

50—What is the cause of pounding in range boilers?

51—What should be investigated first when a complaint of an insufficient supply of hot water is registered?

52—How may such a complaint be due to partial stoppage of a water front?

53—When a new range and water front is connected up to an old boiler and such a complaint is made, what should be the course of investigation that should be followed?

54—How is a boiler connected when the side connection is lower than the upper tapping of the water front?

55—What is understood by the term "quick heating connection"?

56—When a connection of this sort is made are the whole contents of the boiler brought to an equal temperature quicker than with the other style?

57—How is a boiler connected with a water front when a door or window intervenes?

58—What is the usual practice in connecting a gas water heater with the kitchen boiler?

59—How can these connections be made to mitigate stoppages through deposit of sediment in the coils?

60—How may a vertical range boiler be connected with the supply and cold water system to provide an ample flow when the pressure of the cold water supply is low?

61—Is any disadvantage generally attributed to the system of supplying cold water to a boiler through the bottom tapping?

62—How may the connections be made to a standard vertical boiler when it is necessary to fit it in a horizontal position?

63—Why is the provision of a vacuum valve on the boiler considered preferable to the common custom of drilling a hole in the feed tube?

64—Why is it better to have all the tappings of a horizontal boiler in the sides instead of the ends?

65—What disadvantage is likely to result from setting a steam heated horizontal boiler at a level very little above the water level of the steam boiler?

66—What is the object of fitting an equalizing pipe to the steam supply and return pipes of the coil?

67—How may careful proportioning of the steam supply and return pipes offset any possible disadvantage of the low level of a steam heated boiler?

CHAPTER 5.

Pages 46 to 59.

68—What is the principal requirement in the piping connections of a boiler set on the floor below the range with which it is connected to ensure satisfactory heating?

69—What is the cause of circulation in a system constructed in this manner?

70—What is the rule generally followed in arriving at the height of the circulating loop?

71—Is there any possibility of emptying the water front accidentally when the boiler is fitted at a lower level than the range?

72—How may the connections to fixtures be made to effectually prevent emptying of the water front or the circulating loop below the highest point?

73—Is it necessary to provide an air cock on a circulating loop when the supply is taken from an attic tank?

74—Should a fixture on the lower floor be supplied from the pipe connecting the boiler with the water front when the range is on the floor below the boiler?

75—Is it of any advantage to connect the boiler by both top and side tappings when the water front is on the floor below it?

76—What is the best way of connecting an additional horizontal boiler to an existing one when extra storage capacity is required?

77—How may two vertical boilers be connected to ensure that the flow will proceed equally from each?

78—How may two boilers be connected so that the supply from one will be larger than that from the other?

79—What should have chief consideration in connecting two boilers to one water back?

80—Should the same type of connection be used when the boilers are far apart as when they are close together?

81—How may equal distribution to the two boilers be promoted by the use of special fittings?

82—What is the best method of connecting two boilers on different floors to one water front?

83—Why should the range farthest away from the boiler be connected to the top tapping?

84—Is there any possibility of retarding the flow from one heater by the flow from the other?

85—What would be the consequences of this retarding of the flow in the water front?

86—How would the connections to two water fronts on one floor and one on a lower floor be made to one boiler on the upper floor?

CHAPTER 6.

Pages 60 to 73.

87—Is there any objection to passing the flow from a water front on one floor through the water front on a floor above and so into the boiler?

88—Is it good practice to connect more than one water front to the side tapping of a boiler?

89—How should a boiler in a basement be connected with a water front on the floor above and a tank heater in the basement?

90—When a circulating pipe is brought down from a water front on the floor above the boiler and the supply to a circulating loop leaves the boiler at the same tapping will satisfactory service be given?

91—How may a fixture be connected with circulating loop so as to prevent accumulation of air in the system?

92—What is the principal fault experienced with the method of water supply necessitating the connection of more than one boiler to a common main hot water supply line?

93—Is there likely to be as much trouble when the boilers are on different floors as when they are on the same level?

94—What advantage is secured by causing the water in the two boilers to be circulated between each other?

95—Can a system using more than one boiler be operated from a cold water supply to one boiler only?

96—How should a system that is to supply two flats be installed so that the water will be available to each house when heated from either boiler?

97—What is likely to take place if an expansion tank or rather a cold water supply tank over which an expansion pipe has been fitted is placed at a level only slightly above the boiler?

98—Is it possible to take the supply from a heated boiler in a basement and introduce it through the cold water feed tube to a boiler on the next floor with any degree of success?

99—How may the flow from two boilers on the same floor be equalized so that a slight superiority in temperature in one boiler would cause a little stronger flow from that one and an equal flow when the temperature is equal?

CHAPTER 7.

Pages 74 to 90.

100—What are the most objectionable features in the ordinary system of hot water distribution?

101—Are any of the objectionable features in a non-circulating system of hot water distribution emphasised when a low pressure of water obtains?

102—What advantage is obtained by using the header or "water table" method of distribution?

103—What is the best method of constructing such a water table?

104—What is the difference between a primary and a secondary system of hot water circulation?

105—Is it advisable to continue a circulating loop to a fixture under the floor and close up to the fixture?

106—What should be done to prevent the accumulation of air in the circulating pipes?

107—How is a system of circulation with independent loops to each fixture or set of fixtures constructed?

108—Does such a system possess any advantage over a simple continuous circulation system?

109—How is water prevented from backing up the return pipes when a faucet is opened on a circulating loop?

110—Are swing checks better than globe checks on a circulating system?

111—Is it possible to install one system of piping to supply hot water or cold water at will?

112—What must be done to obtain circulation to a fixture when a door or window intervenes between the boiler and the fixture?

113—How may a number of apartments be connected so that each will have a separate boiler heated at will from a gas heater or from a common heater placed in the basement?

114—How should such a system be piped to secure a minimum chance of leakage from the circulating pipes?

115—What is the difference between a drop feed or falling circulation system and a rising supply system?

116—What are the principal points of difference between this system and the English intermediate cylinder system?

117—What is the purpose of a water heated towel rail and how should it be connected to the hot water supply system?

118—How would a circulation system be arranged if the boiler was on the same floor as the fixtures?

119—Where should a safety valve be fitted to be most effective?

120—How may water be circulated to fixtures on a level lower than the boiler?

121—How can the insertion of a valve in the circulating system be made to act as a preventative of reversing the circulation?

CHAPTER 8.

Pages 91 to 100.

122—How is the flow of water equalized on the different floors in buildings of great height?

123—What disadvantage is experienced through the carrying of extra heavy pressures at steam coil heated boilers?

124—What is being done in modern buildings to avoid this disadvantage?

125—What is the most common type of circulating system used in office and apartment buildings?

126—Why are control valves fitted at the base of the loops in a drop feed system as well as at the connection with the main distributing pipe on the top floor?

127—How is the expansion on the long stretches of pipe taken up?

128—How are the lateral connections connected back to cause a circulation to be promoted in them when the distance from the drop pipe to the fixture is great?

129—What is understood by the sectional system of hot water supply?

130—How are the proportions of hot water boilers and heaters for a battery of shower baths estimated?

131—How many gallons of water a minute are passed by the average shower head?

132—Should the amount of water that is possible to be passed through the showers simultaneously be taken as the proper amount to base the proportions of the heater and boiler upon?

133—What is the proper method of estimating the resultant temperature from mixing hot and cold water?

134—How may a simple mixing chamber be constructed so as to provide a means of regulating the temperature at the showers independent of the bather?

135—Describe a simple mixing chamber suitable for showers in factory washrooms or other places where strong construction is necessary.

CHAPTER 9.

Pages 101 to 111.

136—What is the purpose of a double boiler?

137—Is the type of double boiler in which one cylinder is placed within the other the only kind in use?

138—Which of the cylinders is connected with the supply from the attic tank?

139—How may a water front be constructed to serve the same purpose as a double boiler?

140—When two separate boilers are used instead of a double boiler how are the ranges connected?

141—Is there any danger of collapse of the boiler shell of a double boiler when one or other of the boilers is emptied?

142—How should the sediment cocks be placed to obviate the formation of a vacuum and invite a condition that would cause collapse of the boiler?

143—How are the two systems connected so that the supply to the fixtures may be switched from the street to the tank supply if desired?

144—Why is it recommended that a check valve be fitted on the cold water supply pipe from street mains when a double boiler system is used?

CHAPTER 10.

Pages 112 to 127.

145—How may the efficiency of a gas water heater be calculated?

146—What is the average heating value of illuminating gas?

147—How is the amount of gas required to heat a certain volume of water estimated?

148—What is the average efficiency of the ordinary gas water heater?

149—Does a tank heater of the circulating type show as high efficiency as an automatic instantaneous heater?

150—Why are copper coil heaters considered preferable to those using hollow discs as heating surface?

151—What are the principal requirements of a good gas water heater?

152—What is the difference between the internal and external type of thermostatic valve on gas heaters?

153—What is a bath heater of the non-contact type?

154—How should this type of heater be erected so as to supply more than one fixture?

155—Where should the vent pipe that carries the products of combustion be connected to?

156—How should the water connections of a kitchen boiler heater be made when it is feared that sediment may choke the coils?

157—Describe the general construction of a thermostatic gas valve for a water heater?

158—How are the valves on the type of water heater which comes into service on the opening of a faucet operated?

159—Why is it advisable to have the pipe sizes as small as is possible to give a satisfactory flow at the fixtures?

160—What is the advantage of having a heater equipped with thermostatic as well as pressure operated valves?

161—What is the best type of control to use on gas heaters supplying water to large institutions through a storage tank?

162—Why are the doors of the heaters mounted on spring hinges?

163—What are the six principal rules to observe in the installation of gas water heaters?

164—What evidence will be given by a heater of the choking of the heating coils by sediment?

165—How should a continuous flow connection with a water front in a coal range and a gas heater be made?

166—How may a kitchen boiler heater and a boiler be installed to occupy the smallest possible space?

CHAPTER 11.

Pages 128 to 142.

167—Is more heat transmitted from steam to water passing continuously through a tank than to water at rest within the tank?

168—What is the usual allowance of heating surface in steam coils per gallon capacity of storage tank?

169—Why should safety valves be fitted to steam heated storage tanks?

170—How are steam coils in storage tanks controlled by thermostatic valves?

171—How are kitchen boilers heated by means of steam?

172—Can a kitchen boiler be heated by steam without the use of coils?

173—How much quicker may water be heated by injecting steam than by heating it by transmission from coils?

174—How many pounds of water may be heated by injecting one pound of steam at 85-lb. pressure if the water be heated from 70 degrees to boiling point?

175—Does it require more time to heat water by steam at low pressures than at high pressure?

176—How is the quantity of steam necessary to heat a given quantity of water to a determined temperature estimated?

177—How many B.t.u. are there in 1 lb. of steam at 5 lb. pressure?

178—What is the value of a Calorie in B.t.u.?

179—How may the noise of condensing steam when heating water by injection be lessened?

180—Describe the construction of a Co-mingler?

181—What is necessary to prevent water finding its way back into the steam boiler when water is heated in a closed tank by injection?

182—How should a steam boiler and coil be connected up to act as an auxiliary to a tank heater using coal for fuel?

183—What types of heaters are in general use in the fire boxes of hot air and steam heating boilers to provide an auxiliary supply of hot water?

184—Where should a coil be placed in a hot water heating boiler to give the best results in heating water for domestic purposes?

185—What is the best kind of pipe to use in making such a coil?

186—What is the heating value of cast iron water heaters when suspended above the fire in the fire box of a steam heating boiler?

187—How much difference is there between cast iron and brass as a heating medium?

188—How many B.t.u. should be allowed per square foot of coil surface when estimating the heating capacity of a brass coil?

CHAPTER 12.

Pages 143 to 161.

189—What are the usual methods adopted to utilize excess heat generated by a kitchen range or tank heater?

190—How may the size of a radiator that can be heated by a water front in a kitchen range be approximately estimated?

191—What difference should be made in connecting a radiator on the same floor as the water heater and in connecting one on the floor above it?

192—How is the air that is relieved in heating the water prevented from collecting in the system when a radiator is connected to a hot water supply system supplied from an overhead tank?

193—Can a radiator be connected so that the hottest water will be delivered there before being stored in the boiler?

194—Is it preferable to use galvanized iron pipe for coils used in warming rooms when connected to a system supplying water for domestic purposes also?

195—If a radiator is used instead of a coil which pattern should be used if it is to be placed on the same floor as the kitchen range?

196—How much ¾ in., 1 in., 1¼ in., 1½ in. and 2 in. pipe is required to equal 1 sq. ft. of heating surface?

197—How can a room be warmed by installing an extra boiler?

198—Can a room be warmed by means of hot air supplied by a kitchen range?

199—How would the piping system necessary to warm several rooms be laid out if the coils or radiators were to be heated from a kitchen range?

200—How should the connections be made to a plate warming closet heated from the domestic hot water supply?

201—How may a check valve be inserted in the return connection of a circulating system so that it will not prevent free circulation yet will close as soon as a faucet is opened and a tendency to draw through the return pipe is shown?

202—What are the usual types of heated towel rails offered by manufacturers and how are they connected?

203—Describe some schemes of utilizing waste heat in heating water for domestic and manufacturing purposes.

204—How may the heat of waste water from plumbing and other fixtures be turned to account in heating water?

205—Describe the construction of a reliable water heating garbage burner.

206—What are the usual methods of heating water in bakeries?

207—Is it advisable to bury a boiler in the sand over a bakers oven?

CHAPTER 13.

Pages 162 to 176.

208—What is the cause of air locking in a hot water distributing system?

209—What causes lead pipe to sag?

210—Can the method of connecting branch pipes to circulating loops cause air locking under any conditions?

211—How may the practice of carrying a pipe to the basement before rising to the fixtures cause stoppage of supply?

212—How may the cause of intermittent flow at fixtures be sometimes attributed to the position of a relief pipe on a hot water system?

213—Can any mechanical device be utilized to allow air to escape from circulating systems under pressure without allowing water to escape?

214—What is the cause of water continually flowing through a relief pipe into the supply tank of a hot water supply system?

215—Has the elevation of the supply tank any bearing on the subject?

216—How much does water increase in volume by heating?

217—Can water be forced out of the boiler through the relief pipe by pressure of steam formed in the water front by overheating?

218—Is there any advantage in fitting an expansion tank to a relief pipe?

219—When may a trap be used on the supply pipe to prevent circulation back from the hot water tank to the cold water tank?

220—Where should a boiler safety valve be fitted?

221—What is the difference between a safety valve and a vacuum valve?

222—On which part of the system should a vacuum valve be placed?

223—What is the cause of the collapse of copper boilers?

224—What can be done to prevent the collapse of copper boilers when the supply is from the street mains?

225—Can collapse of boilers be laid in any case to lack of pressure in the cold water supply?

226—How can the raising of excessive pressure in hot water supply systems where a check valve is fitted on the supply pipe be avoided?

CHAPTER 14.

Pages 177 to 189.

227—What is the first thing that should be investigated when a complaint of insufficient hot water supply is made?

228—Why has the size of the coal used a bearing on the satisfactory service given by a hot water supply system?

229—Is it possible for a water front to be choked partially so that the circulation to the boiler will be retarded without overheating the water in the water front?

230—What effect on the heating of water is made when the boiler feed tube has become corroded and has dropped off?

231—How may small holes in a kitchen boiler caused by corrosion be satisfactorily repaired?

232—What is the cause of a milky appearance in hot water?

233—Has water hammer any detrimental effect on the piping of a hot water system?

234—Can water be short circuited from a boiler through the connections to a gas water heater and how may it be avoided?

235—What is the usual cause of cold water being drawn at the hot water faucet?

236—What is the cause of rusty water being drawn at the fixtures?

237—How may the drawing of rusty water be avoided by a simple expedient?

238—How would a boiler be tapped to insert an extra connection?

239—Describe the apparatus used for heating water in a barber shop?

240—What is the comparative value of lead and brass as metals for the connecting pipes of range boilers and water fronts?

INDEX.

Spotted something that isn't quite right?

Let us know and we'll fix it for future reprints. And as a thank you for your diligence, we'll send you a free Sothis book of your choice.

publisher@sothis.press

This page intentionally left blank

Made in the USA
Coppell, TX
05 November 2026

11327622R00132